Me & My Conditions

By

JOHN L. BISOL

Me & My Conditions

DISCLAIMERS:

This book is not intended as a substitute for the medical advice of physicians.

The reader should regularly consult a physician in matters relating to his/her health and particularly with respect to any symptoms that may require diagnosis or medical attention.

The information provided within this book is for general informational purposes only and is specific to the author's conditions only. The information was up-to-date and correct at the time of publication. All reference links were active/live at the time of publications. Conflicting medical opinions may exist. There are no representations or warranties, express or implied, about the completeness, accuracy, reliability, suitability or availability of the medical diagnosis with respect to the information, contained in this book. Any use of the information contained is at your own risk.

CITED-WORKS COPYRIGHT'S DISCLAIMER:

This document contains copyrighted material from a variety of public web-sites the use of which was not always been specifically authorized by their copyright owner(s). This material is made available in an effort to advance the understanding of the various diagnosed medical conditions discussed herein.

The author believes that this constitutes a 'fair use' of any such copyrighted material as provided for in section 107 of the US Copyright Law. If you wish to use this copyrighted material from these cited-sources, for purposes of your own – that go beyond 'fair use', you must obtain permission from the original copyrights owners.

First Edition

Printed in the United States of America

TABLE OF CONTENTS

FOREWORD

Whenever my wife asks, "What's wrong with you?" I'm not certain if she is addressing my immediate "state of mind" (I say and do a lot of "weird" things) – or if she is genuinely concerned for my health.

Since I will never have a complete grasp on why I do what I sometimes do, I decided that at the very least I could categorize my physical ailments

Got a "tummy-ache?" Feeling a bit "under the weather?" I'll trade you one of my "conditions" for your lone ailment. Your choice!

I've watched the television commercials and learned just how "devastating" a "Touch of Diabetes" can be to a family. I can certainly relate to that – after 33,603+ (and counting) personal insulin injections.

I can (also) empathize with ESRD (End Stage Renal Disease – got that), or CHF (Congestive Heart Failure – almost killed me this year); or Carotid Artery Blockage (Moderate on one side); Aortic Stenosis – that one runs in the family; or the infinite number of Colon related (bowel problems with cramping, etc.) – you name it, I got it; and don't forget the lung issues, I particularly enjoy gasping for breath. (This list is just a summation of the "high points".)

Want more? How about Secondary Focal Glomerular Sclerosis (aka FGS), ever heard of that? (You won't see my picture on a milk bottle asking for donations.) Of course, FGS goes along with my Polycystic Kidney Disease.

At least they cut out my gallbladder a long time ago, when it calcified.

Yet, here I am, banging out a book detailing the (physical) answers to "What's wrong with you?" I'm not exactly sure if it's "one thing." I believe it's a combination of factors, diseases, conditions and syndromes that are all "working me over" – in conjunction with my life clock running out.

Yes, I've got a few things that need watching. Some of the issues presented herein are immediately life-threatening, other things are life-shortening, a few just add to my daily "misery factor," but I can agree that they all combine to make me one miserable S.O.B. on a "bad day."

Let me get a bit philosophical at this point:

In general, I don't blame any one thing or another for my illnesses, this is just the path I must travel. I have learned to savor the "quiet life" and to more intently feel the pulse of the seasons changing. (I also certainly enjoy a good storm, be it rain or snow or hail or wind.)

The majesty of the power of any storm reminds me that I am a humble creature on this planet.

Take the following philosophical examples for consideration:

I was driving to the Pharmacy last week and the sky was as clear as it could be. I thought, what an insignificant creature I am – yet a work of majesty – to be under the vault of a sky that goes into infinity (the very vastness that defines the universe is revealed all creatures), this work of God, has been revealed to me, as a human, in a special way of appreciation.

I didn't write this book out of "anger," (I hear a lot about people who are "angry" because they have been given an illness to bear.)

I didn't write this book out of "self-pity," – I realize that a lesser-blessed person could not endure the hardship.

I didn't write this book to "blame," – anyone or anything. (I hear a lot about people "blaming God" because they have been given an illness to bear. Some people also "blame" their environment – but there are 8 billion reasons to find exceptions.)

Actually, I wrote this to inform, because I am a façade – "I don't look that sick."

End of Philosophy

NOTES:
- I have added my own personal commentary where needed.
- I am NOT a physician, (I only "see" physicians) therefore, I have formatted my personal remarks in Italics.
- All the "web-links" cited were active on the days noted.
- I can only hope that those of you who told me, "…You should write a book about your illnesses…" will rue the day.
- Each Chapter begins with a "New Diagnosis" of my conditions.

"…One must not think that a person who is suffering is not praying. He is offering up his sufferings to God, and many a time he is praying much more truly than one who goes away by himself and meditates his head off, and, if he has squeezed out a few tears, thinks that is prayer…" – St. Teresa of Avila

Chapter 1: **Let's Start with My Lungs**
There is mild bibasilar linear atelectasis/scarring[1]

The human lung is one of the most essential organs in humans, but it also hosts some horrific conditions and diseases. Atelectasis is one of the conditions where the air sacs (alveoli) in the lungs deflate abnormally. In this case, the outcome is a complete or partial collapse of the lung(s). Usually, this complication arises after surgery. The grave implications of atelectasis are immediately apparent as it impairs the amount of oxygen within the body.

Bibasilar atelectasis affects the left and right bottom parts of the lungs.

Basilar atelectasis is the name given to the condition, in which either a part of the lung or the entire lung collapses due to a hindrance. This condition affects both the left and right lungs. It can be chronic or acute, preventing the respiratory exchange of oxygen and dioxide.

Because of the collapsing of the lungs, another condition known as bibasilar scarring occurs in that part of the lung. Another derivation of this condition is bibasilar sub-segmental atelectasis, which results in the compression or disintegration of a part of the lung distal to blocked segmental bronchus.

What Happens in Basilar Atelectasis?

Knowing about the lung is vital when learning about basilar atelectasis. A human body has a pair of lungs, situated on the left and right sides of the chest. Contrary to common sense, both lungs are not identical; in fact, the left lung has only two partitions, compared to the three of the right lung. Both lungs contain tiny air sacs commonly referred to as alveoli. These sacs are filled with blood vessels, and are constantly involved in gas exchange. When the bottom of the lung (right or left) collapses because of an obstruction, it causes gas exchange to halt. This condition is known as basilar atelectasis.

Symptoms: Usually, the symptoms of atelectasis are hard to pin down as they are similar to those of other diseases or conditions and can be mistaken to be signs of something else.

The symptoms of atelectasis are usually:

---

[1] **Bibasilar Atelectasis:** http://www.newhealthguide.org/Bibasilar-Atelectasis.html
08/23/2016

- Difficulty in breathing
- Rapid or shallow breathing (similar to breathing after any strenuous workout)
- Coughing (*Yes, the famous "coughing for no reason"*)
- Low fever

Complications: The complications resulting from atelectasis can be quite disturbing. They include:

- Low blood oxygen (hypoxemia): this happens because atelectasis hampers the amount of oxygen the alveoli receive.
- Pneumonia: one of the reasons for atelectasis is plugs of mucus. The mucus within a lung affected by atelectasis is an attractive hotbed for various bacterial infections. This can lead to pneumonia. (*Every "common cold" or virus I get lasts for months*)
- Lung scarring: after a lung has collapsed, it needs to be re-inflated. Re-inflation, however, at times, fails to heal all the damage and scarring, which in turn may cause bronchiectasis.
- Respiratory failure: Atelectasis covering a small area is usually curable. Having atelectasis as an adult is comparatively better, however, in adults with lung diseases, it can prove to be lethal if it grows to cover enough area.

Causes of Bibasilar Atelectasis: Atelectasis, more often than not, results from blocked airways. Since, anesthesia hampers the regular airflow-pattern of the lungs, causing a change in absorption of gases. This, in turn, plays a part in the collapsing of the alveoli in the lungs. The other reasons for an obstruction in the lungs (obstructive atelectasis) include:

- Foreign body: any foreign object, which gets inhaled rather than ingested, can cause a collapse of the lungs.
- Mucus plug: during surgery, the drugs injected into the patient cause the lungs to have decreased inflation, prompting the mucus to collect in the airways. An ordinary, medical practice of suction of the lungs helps curb this situation; though, the mucus may keep developing afterwards. This is one of the reasons doctors suggest deep breathing exercises during recovery. Mucus plugs are also found in people suffering from cystic fibrosis, and asthma.
- Blood clot: suffering from severe physical trauma can cause blood to spill out inside the lungs, which, sometimes can't be coughed out.

- Narrowing of major airways from disease: diseases, like tuberculosis and fungal infection, can give rise to scarring in the lungs which constricts the airways.
- Tumor in a major airway: since a tumor is already an obstruction, it can cause atelectasis.

Possible Causes of Non-Obstructive Atelectasis Include:

- Injury: Any serious injury can cause the victim to have reduced breathing due to the pain exhibited. This can lead to compression of the lungs, causing atelectasis.
- Pneumothorax. Air from lungs gets into the space between lungs and chest wall, indirectly causing a little or whole of the lung to get damaged.

Treatments of Bibasilar Atelectasis: The treatment is highly dependent on the condition. If only a small area has been affected, no treatment might be required. If there is an outside obstruction, like a tumor, the treatment focuses on removing or shrinking the obstruction.

This condition alone, (probably) will not kill me outright, but it certainly makes me feel miserable at times, prolongs colds and viruses, and can leave me gasping for air. At least I don't have heart problems – oh wait, I do! (A "lot" of heart problems).

Chapter 2: **Moderate Eventration of the Right Hemidiaphragm**

What is the diaphragm? The diaphragm is the major muscle of respiration and separates the thoracic and abdominal cavities. The phrenic nerve is responsible for the contraction of the diaphragm.

What is diaphragmatic eventration?[2] This is a condition in which the diaphragm is placed at a much higher level than it normally is because it is paralyzed and fails to contract. It most commonly is congenital (present at birth) and may result from a phrenic nerve problem or an abnormally thin diaphragm at birth.

What causes diaphragmatic eventration in adults? In adults this condition is caused by an injury to the phrenic nerve or an infection or a cancer in the chest that causes the phrenic nerve to function poorly. Often a viral infection that results in an eventration may go unnoticed and the results may become apparent several years later. One half of the diaphragm, commonly the left side, is affected. As the lower half of the lung fails to expand properly as the diaphragm does not contract it does not take part in the breathing process effectively.

What symptoms do patients with an eventration experience? The patients may experience respiratory symptoms such as breathlessness, cough or chest pain particularly on exertion. In addition, patients may suffer from recurrent pneumonia, bronchitis, or cardiac arrhythmias. They may also experience gastrointestinal complaints resulting from compression of the stomach.

Another situation which, in itself (alone) would be no "big deal", however when it is acting in concert with other lung problems or heart conditions I guess it could make a person more prone to just "breathe."

---

[2] **Eventration of Diaphragm:** http://www.laparoscopyindia.com/eventration-of-diaphragm/   08/23/2016

Chapter 3: **Tiny Hypodense Thyroid Nodules; Bilaterally**

Thyroid Nodules:[3] The term thyroid nodule refers to an abnormal growth of thyroid cells that forms a lump within the thyroid gland. Although the vast majority of thyroid nodules are benign (noncancerous), a small proportion of thyroid nodules do contain thyroid cancer. In order to diagnose and treat thyroid cancer at the earliest stage, most thyroid nodules need some type of evaluation. (NOTE: Hypodense = usually more consistency of fluid like cyst)

What Causes Thyroid Nodules and How Common Are They? We do not know what causes most thyroid nodules but they are extremely common. By age 60, about one-half of all people have a thyroid nodule that can be found either through examination or with imaging. Fortunately, over 90% of such nodules are benign. Hashimoto's thyroiditis, which is the most common cause of hypothyroidism is associated with an increased risk of thyroid nodules. Iodine deficiency, which is very uncommon in the United States, is also known to cause thyroid nodules.

How Is a Thyroid Nodule Evaluated and Diagnosed? Once the nodule is discovered, a doctor will try to determine whether the rest of the thyroid is healthy or whether the entire thyroid gland has been affected by a more general condition such as hyperthyroidism or hypothyroidism. The physician will feel the thyroid to see whether the entire gland is enlarged and whether a single or multiple nodules are present. The initial laboratory tests may include measurement of thyroid hormone (thyroxine, or T4) and thyroid-stimulating hormone (TSH) in the blood sample to determine whether the thyroid is functioning normally. It's usually not possible to determine whether a thyroid nodule is cancerous by physical examination and blood tests alone, the evaluation of the thyroid nodules often includes specialized tests such as thyroid ultrasonography and fine needle biopsy.

Thyroid Ultrasound: Thyroid ultrasound is a key tool for thyroid nodule evaluation. It uses high-frequency sound waves to obtain a picture of the thyroid. This very accurate test can easily determine if a nodule is solid or fluid filled (cystic), and it can determine the precise size of the nodule. Ultrasound can help identify suspicious nodules since

---

[3] **Thyroid Nodules:** http://www.thyroid.org/thyroid-nodules/   08/25/2016

some ultrasound characteristics of thyroid nodules are more frequent in thyroid cancer than in noncancerous nodules. Thyroid ultrasound can identify nodules that are too small to feel during a physical examination. Ultrasound can also be used to accurately guide a needle directly into a nodule when the doctor thinks a fine needle biopsy is needed. Once the initial evaluation is completed, thyroid ultrasound can be used to keep an eye on thyroid nodules that do not require surgery to determine if they are growing or shrinking over time. The ultrasound is a painless test which many doctors may be able to perform in their own office.

Thyroid Fine Needle Aspiration Biopsy (Fna or Fnab):
A fine needle biopsy of a thyroid nodule may sound frightening, but the needle used is very small and a local anesthetic may not even be necessary. This simple procedure is often done in the doctor's office. Sometimes, medications like blood thinners may need to be stopped for a few days before to the procedure. Otherwise, the biopsy does not usually require any other special preparation (no fasting). Patients typically return home or to work after the biopsy without even needing a band aid! For a fine needle biopsy, the doctor will use a very thin needle to withdraw cells from the thyroid nodule. Ordinarily, several samples will be taken from different parts of the nodule to give the doctor the best chance of finding cancerous cells if they are present. The cells are then examined under a microscope by a pathologist.

The report of a thyroid fine needle biopsy will usually indicate one of the following findings:

1. The nodule is benign (noncancerous).
   - This result is obtained in up to 80% of biopsies. The risk of overlooking a cancer when the biopsy is benign is generally less than 3 in 100 tests or 3%. This is even lower when the biopsy is reviewed by an experienced pathologist at a major medical center. Generally, benign thyroid nodules do not need to be removed unless they are causing symptoms like choking or difficulty swallowing. Follow up ultrasound exams are important. Occasionally, another biopsy may be required in the future, especially if the nodule grows over time.

2. The nodule is malignant (cancerous) or suspicious for malignancy.
   - malignant result is obtained in about 5% of biopsies and is most often due to papillary cancer, which is the most common type of thyroid cancer. A suspicious biopsy has a 50-75% risk of cancer in the nodule. These diagnoses require

surgical removal of the thyroid after consultation with an endocrinologist and surgeon.

3. The nodule is indeterminate. This is actually a group of several diagnoses that may occur in up to 20% of cases. An Indeterminate finding means that even though an adequate number of cells was removed during the fine needle biopsy, examination with a microscope cannot reliably classify the result as benign or cancer.

   - The biopsy may be indeterminate because the nodule is described as a Follicular Lesion. These nodules are cancerous 20- 30% of the time. However, the diagnosis can only be made by surgery. Since the odds that the nodule is not a cancer are much better here (70-80%), only the side of the thyroid with the nodule is usually removed. If a cancer is found, the remaining thyroid gland usually must be removed as well. If the surgery confirms that no cancer is present, no additional surgery to "complete" the thyroidectomy is necessary.

   - The biopsy may also be indeterminate because the cells from the nodule have features that cannot be placed in one of the other diagnostic categories. This diagnosis is called atypia, or a follicular lesion of undetermined significance. Diagnoses in this category will contain cancer rarely, so repeat evaluation with FNA or surgical biopsy to remove half of the thyroid containing the nodule is usually recommended.

4. The biopsy may also be non-diagnostic or inadequate. This result is obtained in less than 5% of cases when an ultrasound is used to guide the FNA. This result indicates that not enough cells were obtained to make a diagnosis but is a common result if the nodule is a cyst. These nodules may require reevaluation with second fine needle biopsy, or may need to be removed surgically depending on the clinical judgment of the doctor.

Nuclear Thyroid Scans: Nuclear scanning of the thyroid was frequently done in the past to evaluate thyroid nodules. However, use of thyroid ultrasound and biopsy have proven so accurate and sensitive, nuclear scanning is no longer considered a first-line method of evaluation. Nuclear scanning still has an important role in the evaluation of rare nodules that cause hyperthyroidism. In this situation, the nuclear thyroid scan may suggest that no further evaluation or biopsy is needed. In most other situations, neck ultrasound and biopsy remain the best and most accurate way to evaluate all types of thyroid nodules.

How Are Thyroid Nodules Treated? All thyroid nodules that are found to contain a thyroid cancer, or that are highly suspicious of containing a cancer, should be removed surgically by an experienced thyroid surgeon. Most thyroid cancers are curable and rarely cause life-threatening problems. Thyroid nodules that are benign by FNA or too small to biopsy should still be watched closely with ultrasound examination every 6 to 12 months and annual physical examination by a doctor. Surgery may still be recommended even for a nodule that is benign by FNA if it continues to grow, or develops worrisome features on ultrasound over the course of follow up.

I wait for the day that the doctor decides to stick a needle in my neck looking for "tiny nodules."

I think it is important (again) to realize that one condition, on its own, is no "big deal," however, during my research I came across an interesting tidbit that I will paraphrase:

Thyroid Deficiency and Mental Health[4]

…At least 13 million Americans suffer from thyroid disorders, and in more than 80% of cases, the problem is an underactive thyroid gland – hypothyroidism.
…Sitting at the base of the throat, the thyroid gland produces hormones that regulate basal metabolic rate, the speed at which our bodies burn food for energy.
…The symptoms of hypothyroidism are variable and sometimes hard to pin down. They may include **fatigue**, *sluggishness, cold intolerance,* **weight gain**, *constipation, muscle or joint pain, thin and brittle hair or fingernails, reduced sexual drive,* **high blood pressure, high cholesterol,** *and a slow heart rate. Patients may also have problems with concentration and memory.*
…But findings have been inconsistent, especially in studies with larger numbers of participants.
…In one such survey, Canadian researchers found that the only psychiatric disorder associated with thyroid disease was **social anxiety disorder (social phobia).** *August 2007 update…*

Note my emphasis, not as excuses but interesting pieces of the overall puzzle. (A person either has the pieces or they are missing the pieces…those are my pieces, combined with other pieces.)

---

[4] **Thyroid Deficiency and Mental Health:**
  http://www.health.harvard.edu/diseases-and-conditions/thyroid-deficiency-and-mental-health
  08/28/2016

Chapter 4: **My Liver and Spleen – Hepatosplenomegaly**
I remember looking at a CT Scan with my doctor and he commented on how my enlarged liver and spleen were absolutely preventing my left lung from properly inflating – fortunately I don't have lung or breathing problems...oh wait...I do!

Hepatosplenomegaly:[5] Enlargement of the liver and the spleen. Following is a list of common causes of Hepatosplenomegaly:

- Leukemia
- Lymphoma
- Autoimmune hepatitis
- Biliart atresia
- Bile duct cancer
- Brucellosis
- Budd Chiari syndrome
- Chagas disease
- Carcinoid syndrome
- Spherocytosis
- Cirrhosis
- Hepatitis C
- Chronic fatigue syndrome
- Malaria
- Galactosemia
- Antenatal infections
- Toxoplasmosis
- Endomyocardial fibrosis
- Thrombocytosis

Symptoms of hepatosplenomegaly:[6] There are some symptoms associated with hepatosplenomegaly and are experienced by the affected individuals. Hepatosplenomegaly may also have other symptoms which may differ according to the underlying disease or disorders. The

---

[5] **Hepatosplenomegaly:**
http://www.rightdiagnosis.com/h/hepatosplenomegaly/intro.htm   08/25/2016
[6] **Symptoms of hepatosplenomegaly**
http://symptomstreatment.org/hepatosplenomegaly-symptomscauses-treatment/
08/25/2016

conditions may be nondescript initially, but after diagnosis a doctor may determine the underlying condition.

Here are some of the symptoms:

- Frequent hiccups
- Mass in the abdomen
- Bruising or easy bleeding (*I routinely suffer from nose-bleeds [Epistaxis]*)
- Tremors or fever
- Digestive issues, especially while digesting large amount of food.
- Looseness in the stools
- Queasiness and vomiting
- Paled skin as well as eye (yellowish skin and eye also known as jaundice)

In a more severe condition the symptoms experienced may be different from mild stage of hepatosplenomegaly. Here are some signs that are experienced during serious stage of the condition:

- Consciousness loss for few seconds or even more or confusion
- Fever of higher degree, over 101° Fahrenheit
- Serious stomach ache
- Paled skin and eye which is also known jaundiced

Causes of hepatosplenomegaly: Here are some common factors that may lead to such conditions:

- IM or "Infectious Mononucleosis" is amongst the common causes of hepatosplenomegaly; it is also commonly referred to as glandular fever. This condition is a highly contagious viral infection which can be transmitted through saliva as well as mucus from person to person. The condition may also lead to engorgement of lymph nodes, serious sore throat as well as high fever. (*aka mono – is often called the "kissing disease" – referring to the normal transmission mode.*)

- Acute viral hepatitis is another condition that can be added to the list of factors causing hepatosplenomegaly. It is an inflammatory condition which is caused due to any of the 5 known hepatitis viruses. Hepatitis A and Hepatitis B are considered as the most common virus types that causes such acute viral infection. Patients who are affected with acute viral hepatitis infection as well as liver and spleen engorgement may also experience high fever, reduced or loss of appetite and jaundice.

Here are some other causes of hepatosplenomegaly:

- Congestive failure of heart (*Probably, in my case*)
- Leukemia
- Metastasis of a tumor

- Neuroblastoma
- Malignant hepatoma
- Niemann pick disease a fatal genetic metabolic disorder
- Reye' syndrome
- HFI or Hereditary fructose intolerance
- HUS or Hemolytic Uremic Syndrome

Treatment for hepatosplenomegaly: It should be known that hepatosplenomegaly treatment may be depended on the factor causing the issue. The enlarged size of liver as well as the spleen may return to the normal state once the root cause of the underlying disease is removed. Hence, for better treatment it is essential to seek immediate medical attention to treat hepatosplenomegaly faster.

Don't panic! I offer these words of encouragement:

__Loss of Brain Function - Liver Disease:__[7] Loss of brain function occurs when the liver is unable to remove toxins from the blood. This is called hepatic encephalopathy. This problem may occur suddenly or develop slowly over time.

__Symptoms:__ Symptoms may begin slowly and slowly get worse. They may also begin suddenly and be severe from the start.
Early symptoms may be mild and include:

- *Change in sleep patterns*
- *Changes in thinking*
- *Confusion that is mild*
- *Forgetfulness*
- *Mental fogginess*
- *Personality or mood changes*
- *Poor judgment*
- *Worsening of handwriting or loss of other small hand movements*

More severe symptoms may include:

- *Abnormal movements or shaking of hands or arms*
- *Agitation, excitement, or seizures (occur rarely)*
- *Disorientation*
- *Drowsiness or confusion*
- *Strange behavior or severe personality changes*
- *Slurred speech*

---

[7] https://medlineplus.gov/ency/article/000302.htm   08/28/2016

- *Slowed or sluggish movement*

"...People with hepatic encephalopathy can become unconscious, unresponsive, and possibly enter a coma. People are often not able to care for themselves because of these symptoms..."

Ibid   08/28/2016

AND

...With advancing liver disease, certain toxins may build up in your body and affect your thinking. You may get irritable, drowsy, or confused (this is called hepatic encephalopathy).

http://docplayer.net/15829412-Living-with-advanced-liver-disease.html

08/28/2016

I feel better already...

Chapter 5: **My Kidneys**

The Initial Diagnosis: Following a "Full, Surgical Biopsy"

1989: **Focal Global and Segmental Glomerulosclerosis,** probably **Secondary Focal Glomerulosclerosis.**

Hyaline arteriolosclerosis.

Note: This probably represents a Secondary Focal Glomerulosclerosis because of the enlarged glomeruli, only focal effacement of the epithelial foot processes by EM and the clinical history rather than primary glomerulosclerosis.

A word (or two) about "My Kidney Cysts":

02/18/2006: Right kidney measures 12.2cm in length and the left 14.1cm in length. There are several cysts present. In the right upper pole there is 3cm x 3.2cm cyst and in the lower pole 2.7 x 2.8cm cyst. In the upper pole of the left kidney, there is a 1.5cm x 2.8cm simple cyst. In the lower to mid pole, there is a 3.8 x 3.4cm septated cyst. The lower portion of the septation appears to have another septation within it and it appears complex in nature.

11/30/2012: There is a small mildly complex right renal upper pole cyst with wall calcification. There is a bilobed 6.7 x 3.7 x 3.8 cm exophytic right lower pole cyst. There is a 1.6 x 1.3 x 1.6 cm isodense complex cyst versus solid mass arising from the left upper pole.

There is a small cyst in the left midpole. There is a small left lower pole parapelvic cyst. There are a few very small bilateral ren, hypodensities, which are too small to characterize. There is no hydronephrosis or nephrolithiasis.

09/18/2015: There is a cyst in the upper pole of the left kidney measuring 36 x 23 mm with a small hyper-dense lesion in the upper pole of the left kidney measuring 13 x 16 mm (which is unchanged.) Scarring and calcification is seen at the upper pole of the right kidney as well as right renal cysts, the largest measuring 34 x 30.5 mm. There is a thin calcified septation in association with the cyst in the lower pole of the right kidney.

FSGS[8] is a rare disease that attacks the kidney's filtering units (glomeruli) causing serious scarring which leads to permanent kidney damage and even failure. FSGS is one of the causes of a serious condition known as Nephrotic Syndrome.

Each kidney is made up of approximately one million tiny filters called "glomeruli." Much as a coffee filter keeps coffee grounds in, glomeruli filter the blood, taking out the water-like part which becomes urine and leaving the protein in the blood. When glomeruli become damaged or scarred (sclerosis), proteins begin leaking into the urine (proteinuria). The word "focal" is added because in FSGS, only some of the glomeruli filters become scarred. "Segmental" means that only some sections of the glomerulus become scarred, just parts of them.

How is FSGS Diagnosed? FSGS is diagnosed with renal biopsy (when doctors examine a tiny portion of the kidney tissue), however, because only some sections of the glomeruli are affected, the biopsy can sometimes be inconclusive.

What are the Symptoms of FSGS? Many people with FSGS have no symptoms at all. When symptoms are present the most common include:

- Proteinuria – Large amounts of protein "spilling" into the urine
- Edema – Swelling in parts of the body, most noticeable around the eyes, hands and feet, and abdomen which causes sudden weight gain.
- Low Blood Albumin Levels because the kidneys are removing albumin instead of returning it to the blood
- High Cholesterol in some cases
- High Blood Pressure in some cases and can often be hard to treat

FSGS can also cause abnormal results of creatinine in laboratory tests. Creatinine is measured by taking a blood sample. Everyone has a certain amount of a substance called creatinine floating in his or her blood. This substance is always being produced by healthy muscles and normally the kidneys constantly filter it out and the level of creatinine stays low. But when the filters become damaged, they stop filtering properly and the level of creatinine left in the blood goes up.

---

[8] **FSGS:** http://nephcure.org/livingwithkidneydisease/understanding-glomerular-disease/understanding-fsgs/   08/28/2016

What Causes FSGS? FSGS is usually "idiopathic," which means it arises without a known cause. There are some known genetic causes of FSGS, with new gene variants continually being discovered.

FSGS can be "primary" or "secondary" in nature:
- Primary FSGS means that the disease happened on its own without a known or obvious reason.
- Secondary FSGS means that doctors think the FSGS was caused by, or is associated with, another medical condition that occurred first. It is not always certain how the other conditions caused the FSGS.

Some causes of secondary FSGS include:
- Kidney defects from birth (dysplasia)
- Urine backing up into kidneys (kidney reflux)
- Viruses and blood disorders (such as HIV and sickle cell anemia)
- Autoimmune disorders (such as lupus and HSP)

Who Gets FSGS? More than 5400 patients are diagnosed with FSGS every year, however, this is considered an underestimate because:
- a limited number of biopsies are performed
- the number of FSGS cases are rising more than any other cause of Nephrotic Syndrome.

How is FSGS Treated? Currently there are few FDA approved treatments for FSGS, but usually a steroid called prednisone or prednisolone is given to try and control proteinuria. Proteinuria Treatment aims to decrease the amount of protein lost in the urine. The less protein in the urine, the better the patient will do. In FSGS, even partial remission is important.

A nephrologist may also recommend:
- Medications that suppress the immune system
- Diuretics and low salt diet help to control edema
- A medication that blocks a hormone system called the renin angiotensin system (ACE inhibitor or ARB) to control blood pressure or lower urine protein
- Anticoagulants to prevent blood clots
- Statins to lower the cholesterol level
- Vitamins

Hyaline arteriolosclerosis:[9] also arterial hyalinosis and arteriolar hyalinosis, refers to thickening of the walls of arterioles by the deposits that appear as homogeneous pink hyaline material in routine staining. It is a type of arteriolosclerosis, which refers to hardening of the arteriolar wall.

It is often seen in the context of kidney pathology. In hypertension only the afferent arteriole is affected, while in diabetes mellitus, both the afferent and efferent arteriole are affected. (*I* **am** *diabetic, therefore, BOTH are affected – can't seem to catch a break.*)

Etiology: Lesions reflect leakage of plasma components across vascular endothelium and excessive extracellular matrix production by smooth muscle cells, usually secondary to hypertension.

Hyaline arteriolosclerosis is a major morphologic characteristic of benign nephrosclerosis, in which the arteriolar narrowing causes diffuse impairment of renal blood supply, with loss of nephrons. The narrowing of the lumen can decrease renal blood flow and hence glomerular filtration rate leading to increased renin secretion and a perpetuating cycle with increasing blood pressure and decreasing kidney function.

All this means I am (now) in Stage 4 – ESRD (End Stage Renal Disease), one hop before dialysis. I have also been informed, by the "good doctors" to be aware of "Kidney Failure"; which, in my case, could occur at any time.
At least I only have 19 other things wrong with my "picture" – it could be worse. I could also have Tinnitus loud enough to drive me crazy…oh wait, I do have Tinnitus…

---

[9] **Hyaline arteriolosclerosis:**
https://en.wikipedia.org/wiki/Hyaline_arteriolosclerosis   08/28/2016

Chapter 6: **My Colon & Associated Polyps**

From: CT Scan 11/30/2012:

- Moderate colonic diverticulosis.
- Nonspecific but non-obstructive mildly distended right abdominal small bowel loops.
- There is suboptimal oral opacification with no oral contrast in the distal small bowel or in the colon.
- There are a few mildly distended nonspecific small bowel loops in the right abdomen. Obstruction is not suggested.
- The colon is redundant and contains mild to moderate fecal material. There are multiple colonic diverticula. There is no significant mural thickening. The appendix is not visualized with certainty, but there are no secondary signs for acute appendicitis.
- There is moderate distention of the ascending and transverse colon with less significant distention of the descending colon.
- Extensive sigmoid colon diverticulosis is seen and there is mild to moderate diverticulitis manifested by wall thickening and fat stranding with inflammatory changes in the left lower quadrant. Evaluation of the descending and sigmoid colon is limited by under distention and the presence of a mucosal abnormality at this level cannot be excluded.
- On the prone images, there is good distention of the rectum which appears unremarkable.
- Proximal sigmoid is within normal limits but the relatively large segment of sigmoid which contains multiple diverticula is not well distended which could be related to circular muscle hypertrophy.
- There is limited evaluation of the mucosa at this level.

Time to take a "gut-wrenching," magical mystery tour through my colon...

What is diverticulosis?[10]: Diverticulosis is a condition that develops when pouches form in the wall of the colon. These pouches are usually very small (5 to 10 millimeters) in diameter but can be larger.

---

[10] **Diverticulosis:**
http://www.webmd.com/digestive-disorders/tc/diverticulosis-topic-overview
08/28/2016

In diverticulosis, the pouches in the colon wall do not cause symptoms. Diverticulosis may not be discovered unless symptoms occur, such as in painful diverticular disease or in diverticulitis. As many as 80 out of 100 people who have diverticulosis never get diverticulitis.1 In many cases, diverticulosis is discovered only when tests are done to find the cause of a different medical problem or during a screening exam.

What causes diverticulosis? The reason pouches (diverticula) form in the colon wall is not completely understood. Doctors think diverticula form when high pressure inside the colon pushes against weak spots in the colon wall.

Normally, a diet with adequate fiber (also called roughage) produces stool that is bulky and can move easily through the colon. If a diet is low in fiber, the colon must exert more pressure than usual to move small, hard stool. A low-fiber diet also can increase the time stool remains in the bowel, adding to the high pressure.

Pouches may form when the high pressure pushes against weak spots in the colon where blood vessels pass through the muscle layer of the bowel wall to supply blood to the inner wall.

What are the symptoms? Most people don't have symptoms. You may have had diverticulosis for years by the time symptoms occur (if they do). Over time, some people get an infection in the pouches (diverticulitis). A doctor may use the term painful diverticular disease. It's likely that painful diverticular disease is caused by irritable bowel syndrome (IBS). Symptoms include diarrhea and cramping abdominal (belly) pain, with no fever or other sign of an infection.

Moderate Diverticulitis:[11] Inflammation puts the "itis" into diverticulitis, which is the most common complication of diverticular disease. The bacteria that are packed into feces by the hundreds of millions are responsible for the inflammation, but doctors don't fully understand why some diverticula become infected and inflamed while many do not. A current theory holds that the wall of the diverticular sac

---

[11] **Moderate Diverticulitis:**
  http://www.health.harvard.edu/diseases-and-conditions/diverticular-disease-of-the-colon
  08/28/2016

becomes eroded by pressure, trapped fecal material, or both. If the damage is severe enough, a tiny perforation develops in the wall of the sac, allowing bacteria to infect the surrounding tissues. In most cases, the body's immune system is able to contain the infection, confining it to a small area on the outside of the colon. In other cases, though, the infection enlarges to become a larger abscess, or it extends to the entire lining of the abdomen, a critical complication called peritonitis.

Pain is the major symptom. Because diverticulosis typically occurs in the sigmoid colon, the pain is usually most pronounced in the lower left part of the abdomen, but other areas may be involved. Fever is also very common, sometimes accompanied by chills. If the inflamed sigmoid is up against the bladder, a man may develop enough urinary urgency, frequency, and discomfort to mimic prostatitis or a bladder infection. Other symptoms may include nausea, loss of appetite, and fatigue. Some patients have constipation, others diarrhea.

Diagnosis: A physician's exam may reveal tenderness over the inflamed tissues, typically in the lower left abdomen; less often, the doctor may feel swelling. As in other infections, the white blood cell counts are usually elevated. But because these findings are non-specific, further testing is required to establish the diagnosis. The best test is a CT scan of the abdomen, ideally performed after the patient receives contrast material both by mouth and intravenously. And a month or two later, after treatment has quieted things down, the patient should have a colonoscopy, both to evaluate the diverticular disease and to be sure that no other abnormalities are lurking.

Therapy: Since bacteria are responsible for the inflammation, antibiotics are the cornerstone of treatment. And because the colon harbors so many bacterial species, doctors must prescribe treatment that will target a broad range of bacteria, including *Bacteroides* and other anaerobic bacteria that grow best without oxygen, as well as *E. coli* and other aerobic (oxygen-requiring) microbes. A common approach is to prescribe metronidazole (Flagyl, generic) for the anaerobes along with ciprofloxacin (Cipro, generic) or trimethoprim-sulfamethoxazole (Bactrim, generic) for the aerobes. Amoxicillin–clavulanic acid (Augmentin) is effective against both types of bacteria and is a good alternative. Needless to say, there are many variations on the theme, and doctors must always take their patients' allergies and general health into consideration when they prescribe antibiotics.

Patients with mild-to-moderate diverticulitis can take their antibiotics in pill form at home, but patients with severe inflammation or complications (see below) should receive intravenous (IV) antibiotics in the hospital, and then finish up with pills at home. In most cases, seven to 10 days of antibiotics will do the trick.

Bowel rest is also important for acute diverticulitis. For home treatment, that means sticking to a diet of clear liquids for a few days, then gradually adding soft solids and moving to a more normal diet over a week or two. Intravenous fluids can sustain hospitalized patients until they are well enough to switch to clear liquids en route to a full diet.

Because diverticulitis tends to recur, prevention is always part of the treatment plan. And for men with any form of colonic diverticular disease, that means a high-fiber diet.

Complications: Ordinary diverticulitis is bad enough, but its complications can be life-threatening. The most common complications include:

Abscess formation. An abscess is a walled-off collection of bacteria and white blood cells – pus. Diverticulitis always involves bacteria and inflammation, but if the body can't confine the process to the wall of the colon immediately adjacent to the perforated diverticulum, a larger abscess forms.

Patients with abscesses tend to be sicker than those with uncomplicated diverticulitis, and they have higher temperatures, more pain, and higher white blood cell counts. Treatment involves antibiotics and bowel rest, but it also requires drainage of the abscess. In many cases, specially trained interventional radiologists can accomplish that by using CT imagery to guide a thin plastic catheter through the skin into the abscess, allowing the pus to drain out. In most cases, the catheter stays in place for several days or until the drainage stops, while the patient continues to receive antibiotics and fluids. Sometimes, though, open surgery is required.

Peritonitis. Although an abscess requires aggressive treatment, it represents a partial success for the body's infection defense apparatus, since the infection is confined to a small area. If that containment fails, infection spreads to the entire lining of the abdomen. Patients are critically ill with high fever, severe abdominal pain, and often low blood pressure. Prompt surgery and powerful antibiotics are required.

Fistula formation. In diverticulitis, the infection can burrow into nearby tissues, such as another part of the intestinal tract, the urinary bladder, or the skin. This complication is less common than abscess formation and less urgent than peritonitis, but it does require both surgery and antibiotics.

Stricture formation. It's another uncommon complication that can develop from recurrent bouts of diverticulitis. In response to repeated inflammation, a portion of the colon becomes scarred and narrowed. Doctors call such narrowing a stricture, and they must call on surgeons to correct the problem so fecal material can pass through without obstruction.

Diverticulitis – Surgery: Most patients with uncomplicated diverticulitis respond well to antibiotics and bowel rest. The majority of patients with abscesses do well with drainage through a catheter, but patients with severe diverticulitis or threatening complications require surgery. Here are some typical indications for surgery:
- Severe diverticulitis that does not respond to medical treatment
- Diverticulitis in patients with impaired immune systems
- Diverticulitis that recurs despite a high-fiber diet
- Abscesses that cannot be drained with a catheter
- Peritonitis, fistula formation, or obstruction
- Strong suspicion of cancer.

The timing and type of operation depend on the patient's individual circumstances. One traditional approach involves two separate operations, the first to remove the disease and divert the intestinal contents to a colostomy bag on the skin, and the second, several months later, to hook the colon and rectum back together. In some cases, this can be accomplished with less-invasive laparoscopic surgery, and in milder cases, one operation may suffice.

Diverticular Bleeding: Diverticulitis is one main complication of diverticular disease of the colon. The other is diverticular bleeding. It occurs when a diverticulum erodes into the penetrating artery at its base. Because acute inflammation is absent, patients with diverticular bleeding don't have pain or fever.

The most common symptom is painless rectal bleeding. Since diverticular bleeding occurs in the colon, it produces bright red or maroon bowel movements. (In contrast, when bleeding occurs in the stomach, the blood is partially digested as it passes through the intestinal tract, so it appears as black, tar-like bowel movements).

In most patients, the bleeding is mild, and it usually stops on its own with bowel rest. But brisk bleeding is a life-threatening emergency. It requires expert hospital care with blood transfusions and IV fluids. It also requires aggressive attempts to locate the site of bleeding and to stop it. Several techniques are available; most experts recommend colonoscopy (doctors can see the bleeding artery through the scope and cauterize or clip it to stop the bleeding) or angiography (doctors thread a catheter into the artery that supplies blood to the colon, inject dye to see the bleeding artery on x-rays, and then inject medication to constrict the artery and stop the bleeding). If neither approach stops the bleeding, surgery may be needed.

Updated: December 2, 2015   Originally published: August 2010

Circular Muscle Hypertrophy:[12] The sigmoid colon affected by diverticular disease appears shortened and thickened. The muscular abnormality is the most important and consistent feature. There is gross thickening of both the longitudinal and circular muscles of the colon, and progressive elastosis of the taeniae coli. This muscular abnormality often precedes the development of diverticulosis and occurs predominantly in the sigmoid colon. The muscle of the sigmoid colon and recto-sigmoid is different from that of the more proximal colon, in that it is thicker and more prone to spasm. The colonic mucosa is pleated, with a saccular appearance. Narrowing of the lumen is due to muscular hypertrophy, redundant mucosal folds and pericolic fibrosis.

In classic situations, diverticula with associated muscular hypertrophy occur predominantly on the left side of the colon and are characterized by inflammation and perforative complications. There appears to be another kind of diverticular disease that is present throughout the entire

---

[12] **Circular Muscle Hypertrophy:**
  http://www.surgwiki.com/wiki/Diverticular_disease_of_the_colon   08/28/2016

colon without associated muscle abnormality. This latter group tends to occur in younger patients and may be due to a connective tissue abnormality that allows development of diverticula. Bleeding as a complication is more common in this atypical group.

Associated Polyps:[13] A colon polyp is a small clump of cells that forms on the lining of the colon. Most colon polyps are harmless. But over time, some colon polyps can develop into colon cancer, which is often fatal when found in its later stages.

Anyone can develop colon polyps. You're at higher risk if the patient is 50 or older, are overweight or a smoker, or have a personal or family history of colon polyps or colon cancer.

Colon polyps often don't cause symptoms. It's important to have regular screening tests, such as colonoscopy, because colon polyps found in the early stages can usually be removed safely and completely. The best prevention for colon cancer is regular screening for polyps.

There are several types of colon polyps, including:

- **Adenomatous.** About two-thirds of all polyps are adenomatous. Only a small percentage of them actually become cancerous. But nearly all malignant polyps are adenomatous.

> *BULLITEN! In January 2016 I had Five (5) Adenomatous polyps removed. Three (3) were identified as exhibiting "severe dysplasia" – precancerous, and two were large adenomatous types.*
> *Prior to that procedure, I had a bluish-red, smooth, adenomatous polyp (a real "biggie") 2 x 1.8 x 1.8 cm. covered by focally ulcerated mucosa; removed in 1994.*

- **Serrated.** Depending on their size and location in the colon, serrated polyps may become cancerous. Small serrated polyps in the lower colon, also known as hyperplastic polyps, are rarely malignant. Larger serrated polyps – which are typically flat (sessile), difficult to detect and located in the upper colon – are precancerous.
- **Inflammatory.** These polyps may follow a bout of ulcerative colitis or Crohn's disease of the colon. Although the polyps themselves are not a significant threat, having ulcerative colitis or

---

[13] Associated Polyps:
http://www.mayoclinic.org/diseases-conditions/colon polyps/basics/definition/con-20031957   08/29/2016

Crohn's disease of the colon increases the overall risk of colon cancer.

Colon polyps often cause no symptoms. But some people with colon polyps experience:

- Rectal bleeding. This can be a sign of colon polyps or cancer or other conditions, such as hemorrhoids or minor tears in the anus.

- Change in stool color. Blood can show up as red streaks in the stool or make stool appear black. A change in color may also be caused by foods, medications and supplements.

- Change in bowel habits. Constipation or diarrhea that lasts longer than a week may indicate the presence of a large colon polyp. But a number of other conditions can also cause changes in bowel habits.

Since I have NO consistent "Bowel Habits" – (I have gone as much as six times in one day to as little as once every other day), it is virtually impossible for me to predict anything that has to do with my colon health. P.S. Don't even mention the word "fiber" around me…excuse me I have to poop.

- Pain, nausea or vomiting. A large colon polyp can partially obstruct the bowel, leading to crampy abdominal pain, nausea and vomiting.

If this were an indicator, I would tell you I have large colon polyps every day.

- Iron deficiency anemia. Bleeding from polyps can occur slowly over time, without visible blood in the stool. Chronic bleeding robs the body of the iron needed to produce the substance that allows red blood cells to carry oxygen to the body (hemoglobin). The result is iron deficiency anemia, which can make the patient feel tired and short of breath.

My colon is "wrecked", and when I "start going" I can't stop…at least I can exclude Kale from my diet.

Chapter 7: **My Carotid Artery**

06/25/2014: **Carotid Imaging Bilateral**

The right common carotid, carotid bulb and the proximal internal and external carotid arteries are visualized. There is probable mild plaque formation in the right carotid bulb. Doppler tones and color-flow are obtained in all vessels with peak velocity of 85/13 in the common carotid artery, 66/6 in the internal carotid artery and peak systolic velocity of 80 in the external carotid artery.

Carotid arteries[14] are the blood vessels that supply blood between the heart and the brain. Over time, plaque buildup obstructs the flow of blood through arteries. Potential outcomes include stroke and heart attack.

However, the amount and type of plaque influence potential outcomes. Soft plaque is most often the cause of a heart attack. Like a domino effect, soft plaque, which is more likely to tear off, can lead to blood clots, blockages, stroke and heart attack. Mild or moderate plaque in the arteries does not require a special procedure.

Eliminating risk factors such as smoking and high blood pressure may also be effective forms of preventing future stroke or heart attack caused by plaque buildup.

OK! Really important to keep those arteries open! No telling when I might have a Cerebrovascular Accident.

Cerebro-Vascular Accident:[15] *The sudden death of some brain cells due to lack of oxygen when the blood flow to the brain is impaired by blockage or rupture of an artery to the brain. A CVA is also referred to as a stroke.*

Symptoms of a stroke depend on the area of the brain affected. The most common symptom is weakness or paralysis of one side of the body with partial or complete loss of voluntary movement or sensation in a leg or arm. There can be speech problems and

---

[14] **Carotid Arteries:**
   https://www.reference.com/health/mild-moderate-hard-plaque-carotid-arteries-mean-ee24e57e729e3176?qo=cdpArticles   08/28/2016

[15] **CVA:**
   http://www.medicinenet.com/script/main/art.asp?articlekey=2676   08/28/2016

weak face muscles, causing drooling. Numbness or tingling is very common. A stroke involving the base of the brain can affect balance, vision, swallowing, breathing and even unconsciousness.

A stroke is a medical emergency. Anyone suspected of having a stroke should be taken immediately to a medical facility for diagnosis and treatment.

The causes of stroke: An artery to the brain may be blocked by a clot (thrombosis) which typically occurs in a blood vessel that has previously been narrowed due to atherosclerosis ("hardening of the artery"). When a blood clot or a piece of an atherosclerotic plaque (a cholesterol and calcium deposit on the wall of the artery) breaks loose, it can travel through the circulation and lodge in an artery of the brain, plugging it up and stopping the flow of blood; this is referred to as an embolic stroke. A blood clot can form in a chamber of the heart when the heart beats irregularly, as in atrial fibrillation; such clots usually stay attached to the inner lining of the heart but they may break off, travel through the blood stream, form a plug (embolus) in a brain artery and cause a stroke. A cerebral hemorrhage (bleeding in the brain), as from an aneurysm (a widening and weakening) of a blood vessel in the brain, also causes stroke.

Chapter 8: **My Heart**

Before we get into this, let me explain that my Paternal Grandfather died of CHF (Congestive Heart Failure) at age 74 years. My own father died of a torn Aorta caused by CHF at age 74 Years – knowing that he needed an Aortic Valve replacement. I always have had a heart "murmur", but so do a lot of people. Unfortunately, I have (already) suffered from Acute Decompensated CHF this year (2016) and my Aortic Valve has sclerosed to 0.8 cm^2 – Normal Area 2.5-4.5 cm^2.

Moderate Stenosis	*0.75-1.0 cm^2*
Severe Stenosis	*< 0.75 cm^2*

Echocardiogram Reports: Combined
Dates of Study: 08/03/2010; 02/23/2015; 10/20/2015; 05/14/2016

M Mode Measurements	Normal Range	Echo Date		
		02/2015	*10/2015*	*05/2016*
Left atrium	(2.5 – 4.0) cm	4.6 cm	5.1 cm	5.5 cm
Aortic Root	(2.1 – 3.7) cm	3.7 cm	3.2 cm	3.0 cm
LV End Diastolic	(4.0 – 5.8) cm	4.1 cm	5.1 cm	5.9 cm
LV End Systolic	(2.2 – 3.9) cm	2.8 cm	2.8 cm	3.9 cm
Intraventricular Septum	(0.6 – 1.1) cm	1.4 cm	1.8 cm	1.4 cm
LV Posterior Wall	(0.6 – 1.1) cm	1.4 cm	1.8 cm	1.4 cm
LVOT Diameter	(1.8 – 2.2) cm	2.0 cm	1.8 cm	1.6 cm
Ascending Aorta	(3.7 – 4.2) cm	4.3 cm	4.0 cm	4.0 cm
Left Ventricular Ejection Fraction	(50 – 80) %	70%	60 %	60%
Left Ventricular Wall Thickness	(0.6 – 1.1) cm	1.1/1.2cm	1.4 cm	1.4 cm
Aortic Valve Area	(3.0 – 4.0) cm^2	1.1 cm^2	1.0 cm^2	0.8 cm^2
Left Atrium	(3.0 – 4.0) cm	4.1 cm	4.6 cm	5.7 cm
Ascending Aorta	(2.1 – 2.8) cm	4.2 cm	4.0 cm	4.0 cm

2-D Echo

- The left atrium appears dilated.
- The left ventricle appears normal in size.
- Left ventricular wall thickness is normal. (02/23/2015)

- Left ventricular wall thickness is significantly hypertrophied. (10/20/2015)
- The right ventricle appears normal in size. (02/23/2015)
- The right ventricle chamber size is possibly dilated. (10/20/2015)
- The Aortic root is sclerotic & normal
- The ascending aorta is mildly dilated.
- No pericardial effusion is present.
- The ascending aorta is 4.2 cm. (02/23/2015)
- The ascending aorta is 3.0 cm. (10/20/2015)

Left Ventricular Wall Motion

- Left ventricular wall motion is probably normal.
- Overall left ventricular systolic function is preserved.

Mitral Valve

- The leaflets are thickened without restriction of motion.
- Mild mitral annular calcification is present.
- The mitral valve has mild regurgitation.
- The mitral valve is mildly thickened with mild mitral annular calcification. (05/14/2015)

Aortic Valve

- The Aortic valve is thickened & calcified with restriction of motion.
- The Aortic valve has moderate stenosis & mild regurgitation.
- The peak gradient is 68 mm Hg., the mean gradient is 43 mm Hg & the Aortic valve area is 1.0 cm^2. (02/23/2015)
- Peak/mean gradients of 74/41 mmHg & a calculated valve area 1.0 sq. cm^2. (10/20/2015)
- The pressure half-time is 477 milliseconds
- The peak Aortic velocity is 4.1 m/sec measured at the apex location.
- The left ventricular outflow tract (LVOT) velocity is 1.3 m/sec.

- Peak Aortic velocity is approximately 4.6m/sec with corresponding peak and mean gradient approximately 84 & 52mm Hg, respectively, calculated Aortic valve area approximately 0.8 cm^2 (05/14/2015)
- The Aortic valve is thickened and sclerotic with reduced

separation. (05/14/2015)
- Demonstrates mild to moderate regurgitation. (05/14/2015)
- Severe Aortic stenosis. (05/14/2015)

Tricuspid Valve
- The tricuspid valve is normal.
- The tricuspid valve has mild regurgitation.
- The estimated PA pressure is: 34 mmHg.
- Demonstrates moderate regurgitation (05/14/2016)

Pulmonic Valve
- The pulmonic valve demonstrates mild regurgitation (05/14/2016)

SUMMARY: All Previous Scans
1. The left ventricle is of normal size with moderate concentric hypertrophy & preserved global ejection traction of 60%.
2. Left ventricular segmental wall motion appears to be normal but not all segments are well visualized.
3. There is a top normal Aortic root & a mildly dilated ascending aorta.
4. The mitral leaflets are thickened with mild regurgitation.
5. The Aortic valve is thickened & calcified with moderate stenosis (AVA 1.0 cm^2) & mild regurgitation.
6. The tricuspid valve has mild regurgitation.
7. The estimated systolic pulmonary artery pressure is borderline at 34 mmHg.
8. Since the previous echo study dated 08/03/2010 ejection fraction was not significantly changed from the previous 55%.
9. There was mild/moderate Aortic stenosis – peak/mean gradients of 46/27 mmHg & a valve area of was 1.6 cm sq. 08/03/2010
10. There is moderate Aortic stenosis – peak/mean gradients of 68/43 mmHg (02/23/2015) & the valve area has progressed to end the valve area is now 1.0 cm2.
11. Previous left ventricular wall thickness between 1.2 & 1.3 cm. 02/23/2015 Now 1.4 cm 10/20/2015.
12. The left atrium has increased from 4.6 cm to 5.7 cm.
13. The left atrium is dilated.

SUMMARY: 05/14/2016:
1. Technically difficult window.
2. Mild to moderate concentric left ventricular hypertrophy.

3. Preserved left ventricular systolic function.
4. Estimated LV ejection fraction is approximately 60%.
5. Within limits of the study there are no obvious segmental findings.
6. RV systolic function appears to be preserved.
7. Mild four-chamber cardiac enlargement.
8. Aortic root appears mildly dilated.
9. Nonspecific mitral thickening with mild mitral annular calcification.
10. Mild mitral insufficiency.
11. Moderate tricuspid regurgitation.
12. Mild pulmonary hypertension.
13. Consistent with impaired LV relaxation.
14. Severe calcific Aortic stenosis, as above.
15. Mild to moderate Aortic insufficiency. May be underestimated in light of difficult window.
16. The left atrial chamber size is mildly dilated.
17. The left ventricle chamber size is mildly dilated.
18. Left ventricular wall thickness is mild to moderately increased concentrically.
19. The right atrial chamber size is mildly dilated.
20. The right ventricle chamber size is mildly dilated.
21. The right ventricular free wall function is normal.
22. The Aortic root is mildly dilated.

On 05/12/2016 family members took me to the emergency room – after a night of being unable to breath. The following is my "diagnosis":

05/12/2016: Patient presents with **SOB (Shortness of Breath) & Acute Decompensated Congestive Heart Failure (CHF)**.
Emergency room ECG: Sinus tachycardia with premature atrial complexes. Left anterior fascicular block. Right bundle branch block. Minimal voltage criteria for LVH, may be normal variant. Abnormal ECG. Final Chest – 1 View Portable Indication: Shortness of breath. Low lung volumes are noted paired hypoventilatory changes are seen. Increased vascular prominence is observed. A left lung base/retro-cardiac opacity cannot be excluded. No definite pneumothorax is seen. No definite pleural effusion is observed within limitations of exam. Impression: Limited Low lung volumes and hypoventilatory changes. Mild vascular congestion. Minimal left lung base atelectasis or infiltrate not entirely excluded.

Chapter 9: **Missing: My Gallbladder Is Surgically Absent 02/18/1992**

I know – it's "been removed", but even missing things add to the "mystique."

FYI:[16] The gallbladder is a small, pear-shaped, sac-shaped organ located in the upper right abdomen – the area between the chest and hips – below the liver, in which bile is stored after secretion by the liver and before release into the intestine.

What are gallstones? Gallstones are hard particles that develop in the gallbladder. Gallstones can range in size from a grain of sand to a golf ball. The gallbladder can develop a single large gallstone, hundreds of tiny stones, or both small and large stones. Gallstones can cause sudden pain in the upper right abdomen. This pain, called a gallbladder attack or biliary colic, occurs when gallstones block the ducts of the biliary tract.

What is the biliary tract? The biliary tract consists of the gallbladder and the bile ducts. The bile ducts carry bile and other digestive enzymes from the liver and pancreas to the duodenum—the first part of the small intestine.

The liver produces bile – a fluid that carries toxins and waste products out of the body and helps the body digest fats and the fat-soluble vitamins A, D, E, and K. Bile mostly consists of cholesterol, bile salts, and bilirubin. Bilirubin, a reddish-yellow substance, forms when hemoglobin from red blood cells breaks down. Most bilirubin is excreted through bile.

The biliary tract: The bile ducts of the biliary tract include the hepatic ducts, the common bile duct, the pancreatic duct, and the cystic duct. The gallbladder stores bile. Eating, signals the gallbladder to contract and empty bile through the cystic duct and common bile duct into the duodenum to mix with food.

What causes gallstones? Imbalances in the substances that make up bile cause gallstones. Gallstones may form if bile contains too much cholesterol, too much bilirubin, or not enough bile salts. Scientists do not fully understand why these imbalances occur.

---

[16] **Gallbladder:**
https://www.niddk.nih.gov/health-information/health-topics/digestive-diseases/gallstones/Pages/facts.aspx   08/28/2016

Gallstones also may form if the gallbladder does not empty completely or often enough.

The two types of gallstones are cholesterol and pigment stones:

- Cholesterol stones, usually yellow-green in color, consist primarily of hardened cholesterol. In the United States, more than 80 percent of gallstones are cholesterol stones.
- Pigment stones, dark in color, are made of bilirubin.

Who is at risk for gallstones? Certain people have a higher risk of developing gallstones than others:

- Women are more likely to develop gallstones than men. Extra estrogen can increase cholesterol levels in bile and decrease gallbladder contractions, which may cause gallstones to form. Women may have extra estrogen due to pregnancy, hormone replacement therapy, or birth control pills.
- People over age 40 are more likely to develop gallstones than younger people.
- People with a family history of gallstones have a higher risk.
- American Indians have genetic factors that increase the amount of cholesterol in their bile. In fact, American Indians have the highest rate of gallstones in the United States – almost 65 percent of women and 30 percent of men have gallstones.
- Mexican Americans are at higher risk of developing gallstones.

Other factors that affect a person's risk of gallstones include:

- Obesity. People who are obese, especially women, have increased risk of developing gallstones. Obesity increases the amount of cholesterol in bile, which can cause stone formation.
- Rapid weight loss. As the body breaks down fat during prolonged fasting and rapid weight loss, the liver secretes extra cholesterol into bile. Rapid weight loss can also prevent the gallbladder from emptying properly. Low-calorie diets and bariatric surgery – surgery that limits the amount of food a person can eat or digest – lead to rapid weight loss and increased risk of gallstones.
- Diet. Research suggests diets high in calories and refined carbohydrates and low in fiber increase the risk of gallstones.
  o Refined carbohydrates are grains processed to remove bran and germ, which contain nutrients and fiber.

- o Examples of refined carbohydrates include white bread and white rice.
- Certain intestinal diseases. Diseases that affect normal absorption of nutrients, such as Crohn's disease, are associated with gallstones.
- Metabolic syndrome, diabetes, and insulin resistance. These conditions increase the risk of gallstones.
  - o Metabolic syndrome also increases the risk of gallstone complications.
  - o Metabolic syndrome is a group of traits and medical conditions linked to being overweight or obese that puts people at risk for heart disease and type 2 diabetes.

Pigment stones tend to develop in people who have:

- Cirrhosis – a condition in which the liver slowly deteriorates and malfunctions caused by chronic, or long lasting, injury infections in the bile ducts
- Severe hemolytic anemias – conditions in which red blood cells are continuously broken down, such as sickle cell anemia

What are the symptoms and complications of gallstones?

Many people with gallstones do not have symptoms. Gallstones that do not exhibit symptoms are called asymptomatic, or silent, gallstones. Silent gallstones do not interfere with the function of the gallbladder, liver, or pancreas.

If gallstones block the bile ducts, pressure increases in the gallbladder, causing a gallbladder attack. The pain usually lasts from one to several hours. Gallbladder attacks often follow heavy meals, and they usually occur in the evening or during the night.

Gallbladder attacks usually stop when gallstones move and no longer block the bile ducts. However, if any of the bile ducts remain blocked for more than a few hours, complications can occur. Complications include inflammation, or swelling, of the gallbladder and severe damage or infection of the gallbladder, bile ducts, or liver.

A gallstone that becomes lodged in the common bile duct near the duodenum and blocks the pancreatic duct can cause gallstone pancreatitis – inflammation of the pancreas.

Left untreated, blockages of the bile ducts or pancreatic duct can be fatal.

When should a person talk with a health care provider about gallstones?

People who think they have had a gallbladder attack should notify their health care provider. Although these attacks usually resolve as gallstones move, complications can develop if the bile ducts remain blocked.

People with any of the following symptoms during or after a gallbladder attack should see a health care provider immediately:

- abdominal pain lasting more than 5 hours
- nausea and vomiting
- fever – even a low-grade fever – or chills
- yellowish color of the skin or whites of the eyes, called jaundice
- tea-colored urine and light-colored stools

These symptoms may be signs of serious infection or inflammation of the gallbladder, liver, or pancreas.

Gallstone symptoms may be similar to those of other conditions, such as appendicitis, ulcers, pancreatitis, and gastroesophageal reflux disease.

How are gallstones treated? If gallstones are not causing symptoms, treatment is usually not needed. However, if a person has a gallbladder attack or other symptoms, a health care provider will usually recommend treatment. A person may be referred to a gastroenterologist—a doctor who specializes in digestive diseases—for treatment. If a person has had one gallbladder attack, more episodes will likely follow.

The usual treatment for gallstones is surgery to remove the gallbladder. If a person cannot undergo surgery, nonsurgical treatments may be used to dissolve cholesterol gallstones. A health care provider may use ERCP to remove stones in people who cannot undergo surgery or to remove stones from the common bile duct in people who are about to have gallbladder removal surgery.

Surgery: Surgery to remove the gallbladder, called cholecystectomy, is one of the most common operations performed on adults in the United States.

The gallbladder is not an essential organ, which means a person can live normally without a gallbladder. Once the gallbladder is removed, bile flows out of the liver through the hepatic and common bile ducts and directly into the duodenum, instead of being stored in the gallbladder.

Surgeons perform two types of cholecystectomy:

Laparoscopic cholecystectomy: In a laparoscopic cholecystectomy, the surgeon makes several tiny incisions in the abdomen and inserts a laparoscope – a thin tube with a tiny video camera attached. The camera sends a magnified image from inside the body to a video monitor, giving the surgeon a close-up view of organs and tissues. While watching the monitor, the surgeon uses instruments to carefully separate the gallbladder from the liver, bile ducts, and other structures. Then the surgeon removes the gallbladder through one of the small incisions. Patients usually receive general anesthesia.

Most cholecystectomies are performed with laparoscopy. Many laparoscopic cholecystectomies are performed on an outpatient basis, meaning the person is able to go home the same day. Normal physical activity can usually be resumed in about a week.

Open cholecystectomy: An open cholecystectomy is performed when the gallbladder is severely inflamed, infected, or scarred from other operations. In most of these cases, open cholecystectomy is planned from the start. However, a surgeon may perform an open cholecystectomy when problems occur during a laparoscopic cholecystectomy. In these cases, the surgeon must switch to open cholecystectomy as a safety measure for the patient.

To perform an open cholecystectomy, the surgeon creates an incision about 4 to 6 inches long in the abdomen to remove the gallbladder. Patients usually receive general anesthesia. Recovery from open cholecystectomy may require some people to stay in the hospital for up to a week. Normal physical activity can usually be resumed after about a month.

That's just "great!" Why did I have my gallbladder removed? Too many large stones! I had a severe attack of pancreatitis, my body turned "yellow" and it was off to the hospital for me. I had an "Open Cholecystectomy" because my gallbladder was severely inflamed and infected. Anyway, the legend goes that the pathology lab needed a "hammer" to break the stones open for examination…and yes, I was very sick before they took my gallbladder…very bilious to be precise.

Chapter 10: **Prostate – Mildly Enlarged with Punctate Calcifications**

Prostatic calcification[17], or prostate stones, are one cause of chronic prostatitis/chronic pelvic pain syndrome. The stones are common in men. Approximately 75% of middle-aged men have prostatic calcification. The stones themselves do not usually cause symptoms and may be related to an enlarged prostate from benign prostatic hyperplasia (BPH) and are commonly found in men with prostate cancer. They can aggravate lower urinary tract symptoms in some men. Other men do not even know they have them until they are getting screened for something else. Between men with urinary symptoms and men without symptoms, there does not seem to be a difference in the stones' numbers, sizes, or locations in the prostate.

The stones become problematic and can lead to prostatitis if they serve as a source of recurring infection. Even if the patient takes antibiotics that kill the bacteria associated with the stones, the obstructive stones still remain, so the inflammatory process continues. The chronic inflammation can lead to prostatitis symptoms. Inflammation is one of the body's immune system defense mechanisms. While inflammation kills germs, it can also harm other tissues. Inflammation can cause problems with the nervous system, causing pain in the case of prostatitis. Sometimes bacteria continue to live in the stones but it does not show up when cultured because it is sealed off by the stone or a scar.

It is not for certain why the stones form. Some experts say they are from prostatic secretions. Others say the stones tend to be made from ingredients found in urine and not prostate secretions. This kind would form from urine making its way into the prostatic ducts. Anecdotal evidence seems to suggest that taking magnesium and zinc helps to break up these stones and the patient may see "gravel" in semen or urine if this occurs. Other doctors say that there is no dietary change or supplement that will help with stones caused by prostatic secretions; so that issue is controversial.

---

[17] **Punctate Calcifications**:
http://prostatitis.net/causes-of-prostatitis/causes-chronic-prostatitis/prostatic-calcification/   08/25/2016

I can hardly wait for a "flare-up" or perhaps, an infection. Meanwhile, my doctor has prescribed "watchful waiting" which (in lay terms) means I get to "bend over" during most of my visits. Just checking for prostate stones...just checking.

Chapter 11: **Mild Contrast in The Distal Esophagus Suggesting Gastroesophageal Reflux or Stasis – There Is a Very Small Hiatal Hernia**

Gastroesophageal reflux disease[18], or GERD, is a digestive disorder that affects the lower esophageal sphincter (LES), the ring of muscle between the esophagus and stomach. Many people, including pregnant women, suffer from heartburn or acid indigestion caused by GERD. Doctors believe that some people suffer from GERD due to a condition called hiatal hernia. In most cases, GERD can be relieved through diet and lifestyle changes; however, some people may require medication or surgery.

Gastroesophageal refers to the stomach and esophagus. Reflux means to flow back or return. Therefore, gastroesophageal reflux is the return of the stomach's contents back up into the esophagus.
In normal digestion, the lower esophageal sphincter (LES) opens to allow food to pass into the stomach and closes to prevent food and acidic stomach juices from flowing back into the esophagus. Gastroesophageal reflux occurs when the LES is weak or relaxes inappropriately, allowing the stomach's contents to flow up into the esophagus.

What Is the Role of Hiatal Hernia in GERD?
Some doctors believe a hiatal hernia may weaken the LES and increase the risk for gastroesophageal reflux. Hiatal hernia occurs when the upper part of the stomach moves up into the chest through a small opening in the diaphragm (diaphragmatic hiatus). The diaphragm is the muscle separating the abdomen from the chest. Recent studies show that the opening in the diaphragm helps support the lower end of the esophagus. Many people with a hiatal hernia will not have problems with heartburn or reflux. But having a hiatal hernia may allow stomach contents to reflux more easily into the esophagus.

Coughing, vomiting, straining, or sudden physical exertion can cause increased pressure in the abdomen resulting in hiatal hernia. Obesity and pregnancy also contribute to this condition. Many otherwise healthy

---

[18] **Gastroesophageal Reflux:** http://www.webmd.com/heartburn-gerd/guide/reflux-disease-gerd-1#1   08/28/2016

people age 50 and over have a small hiatal hernia. Although considered a condition of middle age, hiatal hernias affect people of all ages.

Hiatal hernias usually do not require treatment. However, treatment may be necessary if the hernia is in danger of becoming strangulated (twisted in a way that cuts off blood supply, called a paraesophageal hernia) or is complicated by severe GERD or esophagitis (inflammation of the esophagus). The doctor may perform surgery to reduce the size of the hernia or to prevent strangulation.

Add to the list: ...after eating I feel like "throwing up." I have discovered(?) that a little Baking Soda in a glass of water gives me enough of a "burp" to kill-off the pressure and reduce the stomach acid. I fondly call this my "Burping Medicine" regimen. Of course, having a "...very small hiatal hernia" is akin to having a small leak in your rowboat, eventually you have to "bail out the water."

Chapter 12: **There Are Small Fat-Containing Bilateral Inguinal Hernias**

An inguinal hernia[19] occurs when tissue pushes through a weak spot in the groin muscle. This causes a bulge in the groin or scrotum. The bulge may hurt or burn. They can sometimes occur on both sides of the pubic area, and if they do, they are called bilateral inguinal hernias.

Most inguinal hernias happen because an opening in the muscle wall does not close as it should before birth. That leaves a weak area in the belly muscle. Pressure on that area can cause tissue to push through and bulge out. A hernia can occur soon after birth or much later in life.

Hernias are more common in men. The main symptom of an inguinal hernia is a bulge in the groin or scrotum. It often feels like a round lump. The bulge may form over a period of weeks or months. Or it may appear all of a sudden after lifting heavy weights, coughing, bending, straining, or laughing. The hernia may be painful, but some hernias cause a bulge without pain.

A hernia also may cause swelling and a feeling of heaviness, tugging, or burning in the area of the hernia. These symptoms may get better when the patient lies down.

Sudden pain, nausea, and vomiting are signs that a part of the intestine may have become trapped in the hernia. (Call the doctor if the patient has a hernia and has these symptoms.)

Surgery is the only treatment and cure for inguinal hernia. Hernia repair is one of the most common surgeries done in the United States. About 750,000 people have hernia repairs each year.

Many doctors recommend surgery to repair a hernia because it prevents strangulation, which occurs when a loop of intestine is trapped tightly in a hernia and the blood supply is cut off, which kills the tissue.

A strangulated hernia requires immediate surgery, although the condition is rare in adults.

Waiting to have surgery does not increase the chance that part of the intestine or abdominal tissue will get stuck in the hernia. Waiting will

---

[19] **Bilateral Inguinal Hernias:** http://www.webmd.com/digestive-disorders/tc/inguinal-hernia-topic-overview   08/28/2106

also not increase the risk for problems, if the patient decides to have surgery later. In some cases, hernias that are small and painless may never need to be repaired.

Yes, my hernia(s) cause swelling and a feeling of heaviness, tugging, or burning in the area of the hernia(s) – That makes me feel "grumpy." These symptoms get better when I lie down, but only in a certain way so as not to initiate my orthopnea[20].

In case you're wondering, I am acutely aware of the following: "…Sudden pain, nausea, and vomiting are signs that a part of the intestine may have become trapped in the hernia. Call the doctor if you have a hernia and have these symptoms…"

---

[20] A form of Dyspnea – (Difficulty in breathing, often associated with lung or heart disease and resulting in shortness of breath. Also called air hunger.) – in which the person can breathe comfortably only when standing or sitting erect;

Chapter 13: **Mild > Moderate Vascular**[21] **Calcifications**

Cardiovascular complications are the leading cause of death in patients with chronic kidney disease (CKD). Vascular calcification[22] is a common complication in CKD, and investigators have demonstrated that the extent and histoanatomic type of vascular calcification are predictors of subsequent vascular mortality. Although research efforts in the past decade have greatly improved our knowledge of the multiple factors and mechanisms involved in vascular calcification in patients with kidney disease, many questions remain unanswered. No longer can we accept the concept that vascular calcification in CKD is a passive process resulting from an elevated calcium-phosphate product. Rather, as a result of the metabolic insults of diabetes, dyslipidemia, oxidative stress, uremia, and hyperphosphatemia, "osteoblast-like" cells form in the vessel wall. These mineralizing cells as well as the recruitment of undifferentiated progenitors to the osteochondrocyte lineage play a critical role in the calcification process.

I really hate those words, "…Vascular mortality." Ugh! If my vessels, especially those that carry blood die, can I be far behind?

---

[21] **Vascular:** relating to, affecting, or consisting of a vessel or vessels, especially those that carry blood.

[22] **Vascular Calcifications:** http://jasn.asnjournals.org/content/20/7/1453.full
08/28/2016

Chapter 14: **Degenerative Changes in The Spine, Sacroiliac Joints, & Hips**

The phrase **"Degenerative Changes" in the spine** refers to osteoarthritis of the **spine**. Osteoarthritis is the most common form of arthritis. Doctors may also refer to it as **degenerative** arthritis or **degenerative** joint disease. Osteoarthritis in the **spine** most commonly occurs in the neck and lower back.[23]

With age, the soft disks that act as cushions between the spine's vertebrae dry out and shrink. This narrows the space between vertebrae, and bone spurs may develop. Gradually, the spine stiffens and loses flexibility. In some cases, bone spurs on the spine can pinch a nerve root – causing pain, weakness or numbness.

Sacroiliac Joints:[24] The sacroiliac joint or SI joint (SIJ) is the joint between the sacrum and the ilium bones of the pelvis, which are connected by strong ligaments. In humans, the sacrum supports the spine and is supported in turn by an ilium on each side. The joint is a strong, weight transferal synovial plane joint with irregular elevations and depressions that produce interlocking of the two bones. The human body has two sacroiliac joints, one on the left and one on the right, that often match each other but are highly variable from person to person. While it is not clear how the pain is caused, it is thought that an alteration in the normal joint motion may be the culprit that causes sacroiliac pain.

Patients who have osteoarthritis of the hip sometimes have problems walking. Diagnosis can be difficult at first. That's because pain can appear in different locations, including the groin, thigh, buttocks, or knee. The pain can be stabbing and sharp or it can be a dull ache, and the hip is often stiff. [25]

No, I cannot "…walk a mile in your shoes…" I can barely wear shoes and walk.

---

[23] **Degenerative Changes:**
http://www.mayoclinic.org/diseases-conditions/osteoarthritis/expert-answers/arthritis/faq-20058457   08/29/2016
[24] http://www.spine-health.com/conditions/sacroiliac-joint-dysfunction/sacroiliac-joint-dysfunction-si-joint-pain   08/29/2016
[25] http://www.webmd.com/osteoarthritis/guide/hip-osteoarthritis-degenerative-arthritis-hip   08/29/2016

Chapter 15: **The Urinary Bladder Is Incompletely Distended**

Urinary retention and incomplete bladder emptying can be caused by an inadequately contractile bladder, urethral sphincter obstruction, or both. There are three areas of the central nervous system that control bladder function: the sacral micturition center, the pontine micturition center, and the cerebral cortex. The sacral micturition center is located in the spinal cord at the sacral (S2-S4) levels and is responsible for bladder contraction. The pontine micturition center is located in the brainstem and appears to play a role in coordinating relaxation of the external sphincter with bladder contractions. The cerebral cortex plays an inhibitory role in relation to the sacral micturition center. Therefore, trauma to either of these three centers may cause difficulty with the control of bladder function. There is a condition known as detrusor hyperreflexia with impaired contractility (DHIC), which refers to overactive bladder symptoms, but the detrusor - (muscle of the bladder wall. The contraction of this muscle causes bladder contraction and voiding). The external sphincter is in synergy with detrusor contraction, but the detrusor is too weak for proper voiding to occur. However, with all that said, the patient would be advised to consult the doctor or gynecologist in order to have the situation clearly explained, and to ask any further questions that the patient may have.

It is an embarrassing situation when the patient has an urge to pass urine, but the patient may not be able to void immediately due to various reasons. The result is that the patient has a distended bladder with mild pain in lower abdomen. Often we all come across such situation mostly while traveling. We only feel relieved after emptying our bladder[26].

However, apart from these normal phenomena that we sometime face, there are many pathological conditions that may cause distended bladder. There may be an urge to pass urine, but the person may not be able to pass urine or if at all, he may void minimal quantity.

---

[26] **Urinary Bladder Is Incompletely Distended:** http://www.simple-remedies.com/health-tips-6/distended-bladder-causes-treatment.html 08/26/2016

This typical presentation is found in retention of urine, which leads to distension of bladder. Distension of bladder is a result of large accumulation of urine.

Causes of Distended Bladder: Bladder distension can occur primarily due to two main reasons:

- If there is a blockage in the pathway of urine expulsion.
- If there is damage to the nerve or the bladder wall that controls urination.

Urinary tract causes: A large sized stone in bladder can block the urine outflow and cause accumulation of urine in bladder. Similarly, prostate enlargement (Benign prostate hypertrophy, BPH) may obstruct the urethral passage and cause retention of urine with bladder distension.

The flow of urine may be completely blocked or partially blocked, depending on the size of the growth. Problems in urethra such as stricture, spasm, infection or an ulcer may lead to partial or complete blockage of urine. Urethra is a passage through which urine flows out of the body. Urethra is connected to urinary bladder.

Neurological: Damage to nerves which control bladder contraction and relaxation to empty the collected urine may be a cause for distended bladder. These nerves can get compressed and damaged as a result of spinal cord compression, spinal cord tumor, or injury to spinal cord. Infection of spinal cord such as pott's disease can also compress the spinal nerves and become a cause for bladder distention.

Distended Urinary Bladder Symptoms:

Pain in lower abdomen: Most cases of distended bladder come with the first complaint of pain in lower abdomen. The lower abdomen may feel swollen when it is palpated. It feels tensed and tender when toughed. This is particularly felt in a thin person. However, pain may not be elicited in old age people or those who are in coma. If the condition is chronic, then there is less pain as compared to acute state.

Difficulty to urinate: Patient may find difficulty to pass urine, even though there is an urge to pass it. There may be partial outflow after great straining. The person may feel unsatisfied and there is always a feeling of bladder fullness.

Prolonged time for urination: the person may find an extended period of time for actual flow of urine to take place after trying to urinate.

In neurological causes there may be associated symptoms such as weakness in muscles, pins and needles in limbs, decreased muscle tone etc.

Treatment Options for Distended Bladder:

- In most cases of acute bladder distension, applying a hot water bag or sitting in a bathtub containing hot water may be effective. (It relieves the congestion of bladder and allows easy voiding of urine.)
- Reassurance and encouragement to pass urine is also valuable during this period.
- Many times hearing or watching the flow of tap water helps to start the urine flow.
- In case if all the home treatment options fail, the last option is to visit a hospital and consult a doctor. In the hospital, they may use catheter (a tube) for removal of accumulated urine in a distended bladder. Full aseptic precautions are taken while introducing the catheter inside the urethra and bladder.
- In case of neurological causes, new implantable devices have come, which stimulate the nerves and muscles of bladder to contract and relax when the need arises.

Let's review...I "over-poop" and "under pee," except when I can't "poop" and can't stop "peeing." Very interesting set of conditions...

Chapter 16: **Asymmetric Gynecomastia Greater on the Left**

Gynecomastia[27] is swelling of the breast tissue in boys or men, caused by an imbalance of the hormones estrogen and testosterone.

Gynecomastia can affect one or both breasts, sometimes unevenly.

Generally, gynecomastia isn't a serious problem, but it can be tough to cope with the condition. Men with gynecomastia sometimes have pain in their breasts.

Gynecomastia may go away on its own. If it persists, medication or surgery may help.

Gynecomastia is triggered by a decrease in the amount of the hormone testosterone compared with estrogen. The cause of this decrease can be conditions that block the effects of or reduce testosterone or a condition that increases the estrogen level. Several things can upset the hormone balance, including the following.

Natural hormone changes: The hormones testosterone and estrogen control the development and maintenance of sex characteristics in both men and women. Testosterone controls male traits, such as muscle mass and body hair. Estrogen controls female traits, including the growth of breasts.

Most people think of estrogen as an exclusively female hormone, but men also produce it – though normally in small quantities. However, male estrogen levels that are too high or are out of balance with testosterone levels can cause gynecomastia.

Medications: A number of medications can cause gynecomastia. These include:
- Anti-androgens used to treat prostate enlargement, prostate cancer and some other conditions.
  - Examples include flutamide, finasteride (Proscar, Propecia) and spironolactone (Aldactone).
- Anabolic steroids and androgens.

---

[27] **Gynecomastia:** http://www.mayoclinic.org/diseasesconditions/gynecomastia/basics/definition/con-20028710   08/28/2016

- AIDS medications.
- Anti-anxiety medications, such as diazepam (Valium).
- Tricyclic antidepressants.
- Antibiotics.
- Ulcer medications, such as cimetidine (Tagamet).
- Cancer treatment (chemotherapy).
- Heart medications, such as digoxin (Lanoxin) and calcium channel blockers.

Health conditions: Several health conditions can cause gynecomastia by affecting the normal balance of hormones. These include:

- Hypogonadism. Any of the conditions that interfere with normal testosterone production, such as Klinefelter syndrome or pituitary insufficiency, can be associated with gynecomastia.
- Tumors. Some tumors, such as those involving the testes, adrenal glands or pituitary gland, can produce hormones that alter the male-female hormone balance.
- Hyperthyroidism. In this condition, the thyroid gland produces too much of the hormone thyroxine.
- Kidney failure. About half the people being treated with regular hemodialysis experience gynecomastia due to hormonal changes.
- Liver failure and cirrhosis. Hormonal fluctuations related to liver problems as well as medications taken for cirrhosis are associated with gynecomastia.
- Malnutrition and starvation. When a body is deprived of adequate nutrition, testosterone levels drop, but estrogen levels remain constant, causing a hormonal imbalance. Gynecomastia can also occur once normal nutrition resumes.

This finding "proves" that I definitely have a "feminine side"!

Chapter 17: **Mild Panniculitis**
(No abscess is seen. There are lower ventral abdominal wall skin thickening and subcutaneous fatty induration[28].)

Panniculitis[29] is a broad term referring to inflammation of the fatty layer underneath the skin. Panniculitis is a group of diseases whose hallmark is inflammation of subcutaneous adipose tissue (the fatty layer under the skin - panniculus adiposus). Symptoms include tender skin nodules, and systemic signs such as weight loss and fatigue

There are many types of panniculitis – with different causes – but the condition generally causes the skin to feel hard and to develop painful red lumps (nodules) or patches (plaques), making it look darker in places.

Panniculitis usually affects the shins and calves, but may spread to the thighs, forearms and chest. It tends to clear up within six weeks, fading like a bruise, usually without scarring.

When the inflammation has settled, a depression in the skin may be left, which can be temporary or permanent.

Many people get recurring bouts of panniculitis.

Other symptoms – As well as skin symptoms, panniculitis may also be associated with:
- fever
- fatigue
- weight loss
- nausea and vomiting
- joint pain

What are the causes? There are a wide range of possible causes of panniculitis, although often the cause is not known. Common causes include:
- an infection – usually a viral or bacterial infection

---

[28] **Induration:** Localized hardening of soft tissue of the body. The area becomes firm, but not as hard as bone. 08/28/2016

[29] **Panniculitis:** http://www.nhs.uk/conditions/panniculitis/Pages/Panniculitis.aspx 08/28/2106

- an inflammatory disease such as Crohn's disease or ulcerative colitis
- medicines, sulphonamides (a group of antibiotics)
- sarcoidosis – a rare disease that causes body cells to form into clumps, called granulomas, in the lungs and skin
- leukemia (cancer of the white blood cells) or lymphoma (cancer of the lymphatic system, part of the immune system)
- Some cases of panniculitis may be caused by the body's immune system mistakenly attacking the fat cells.

Types of panniculitis: The layer of fat underneath the skin is made of lobules (groups of fat cells) held together by connective tissue. Doctors sometimes classify the disease as either:

- 'mostly septal' – the inflammation mostly affects the connective tissue
- 'mostly lobular' – it mostly affects the fat lobules

Some people will also have vasculitis, where the immune system attacks the body's blood vessels. If a blood vessel is inflamed, it can narrow or close off, this can limit, or even prevent, blood flow through the vessel and potentially damage organs.

The most common type of panniculitis is erythema nodosum, which affects the shins. In about half of all cases of erythema nodosum, the cause is unknown.

A similar form of the disease is Weber-Christian disease, also known as idiopathic lobular panniculitis (idiopathic means 'unknown cause'). This most commonly affects the thighs and lower legs of women aged 30-60, and can also cause the non-skin symptoms mentioned above, such as fever and fatigue.

Other types include:

- erythema induratum (nodular vasculitis), which usually affects the calves of young women and is often caused by tuberculosis
- cold panniculitis, which affects areas of skin exposed to the cold – for example, it can affect the cheeks and forehead of infants and children
- subcutaneous sarcoidosis, when the cause is the rare disease sarcoidosis

How is panniculitis treated? Doctors will aim to treat the underlying cause of the panniculitis, if known, and relieve some of the symptoms. While treatment is underway, the patient will be asked to ensure that they get enough rest and to elevate the affected area when possible.

Treatments will vary; for example, if panniculitis is triggered by medication, this medicine should be stopped.

- If the cause is a bacterial infection, the patient will be prescribed anti-inflammatory antibiotics, such as tetracycline, to clear the infection.
- If the cause is sarcoidosis, the patient may not need any medical treatment as the disease often goes away on its own with time (usually a couple of years). Often, simple lifestyle changes, over the counter painkillers, and support bandages are all that is needed to control any flare ups. Read more about managing sarcoidosis.

The following treatments may help to relieve symptoms:

- Joint pain and painful skin nodules can be relieved with anti-inflammatory painkillers (NSAIDs), such as ibuprofen.
- A solution of potassium iodide may help to relieve symptoms – this is thought to have an effect on white blood cells (read more about potassium iodide treatment for skin conditions).
- The inflammation may occasionally be treated with steroid cream, steroid tablets or injections, or immunosuppressants (drugs to weaken the immune system), if the immune system is responsible.

This "theory" about elevating the legs does not apply in my case. I get excruciating pain when my legs are elevated. (Not to mention that I cannot breathe very well in that position – which leads to a panic attack.)

If I take anti-inflammatory painkillers I tend to "bleed like a stuck pig" through my nose.

With a shrug of my shoulders I say… "I am special!"

Chapter 18: **Bloody Nose – Bloody Hell!**

Nosebleeds[30] (aka Epistaxis) are common, and while the cause may be unclear at first, most cases are minor and can be managed from home.

Immediate causes of nosebleeds include trauma to the nose from an injury, deformities inside the nose, inflammation in the nose, or, in rare cases, intranasal tumors. Any of these conditions can cause the surface blood vessels in the nose to bleed.

Sudden and inexplicable nosebleeds may seem scary, but typically they're not.

Underlying Health Conditions:
Liver disease, kidney disease, chronic alcohol consumption, or another underlying health condition can lower the blood's ability to clot and therefore cause the nose to bleed.

Heart conditions like hypertension (high blood pressure) and congestive heart failure can also cause nosebleeds, as can hypertensive crisis – a sudden, rapid increase in blood pressure that may be accompanied by a severe headache, shortness of breath, and anxiety.

Colds, allergies, and frequent nose-blowing can also irritate the lining of the nose, resulting in a nosebleed.

Dry Air:
Dry air from indoor heating or outdoor cold can dry the lining of the nose, causing it to crack and bleed. Using a humidifier while sleeping can help relieve dryness, and nasal sprays are helpful for moistening the nostrils.

Blood-Thinning Medications:
Anticoagulant (blood-thinning) medications, aspirin, and nonsteroidal anti-inflammatory drugs (NSAIDS) used to treat pain can all cause nosebleeds. Because blood clotting is a necessary step in preventing or

---

[30] **Nosebleeds:**
http://www.everydayhealth.com/heart-health/nose-bleed-for-no-reason-here-are-possible-causes-3856.aspx   08/30/2016

stopping a nosebleed, any medication that changes the blood's ability to clot can cause a bloody nose – or make one harder to stop.

Nose Picking or Scratching
Accidental injury to the blood vessels in the nostril from nose picking can cause a nosebleed. This is common in children, but also in adults who are prone to itching or scratching inside their noses.

How to Stop a Bloody Nose at Home
- While sitting and leaning forward, use direct pressure to stop bleeding by pinching the nostrils shut for at least 10 minutes, breathing through the mouth.
- Alternatively, the patient can make a nose-pinching device using tongue depressors and tape.
- If bleeding starts again, use a nasal decongestant spray (such as Afrin) to constrict the blood vessels of the nose, and again apply direct pressure to stop bleeding.
- To prevent another bloody nose, use saline and topical ointments to moisturize inside the nose, but only once bleeding has stopped.

When to Get Help for Nosebleeds
Having more than one nosebleed a week is also a sign that the patient should talk to a doctor. If nosebleeds are recurrent – whether or not the patient is on blood-thinning medications – it's reasonable to seek help from a primary care physician, recurrent nosebleeds may indicate other, more significant medical conditions.

Seek medical attention in an emergency room if the nosebleed lasts longer than a few minutes, or if the patient is unable to stop the bleeding with direct manual pressure.

In truth, I "average" at least three nosebleeds a week. I have gone through some bad times when I have had a "daylong" nosebleed. We have a plastic bag that I use for my bloody cloths – so that they can be separately washed. Most of my nosebleeds last from 10 minutes to one hour (or more).
I (also) get nosebleeds when I chew my food vigorously, although a doctor will tell the patient that is "impossible." Sneeze…nosebleed. Sitting down watching TV…nosebleeds. In the shower…nosebleeds. I think you get the picture. It is embarrassing to say the least and can be debilitating.

Chapter 19: **A "Touch" Of Diabetes** or: *45,000⁺ "finger-sticks," 33,603⁺ personal insulin injections and counting.*

Diabetes[31] is a problem with the body that causes blood glucose (sugar) levels to rise higher than normal. This is also called hyperglycemia. Type 2 diabetes is the most common form of diabetes.

In type 2 diabetes the body does not use insulin properly. This is called insulin resistance. At first, the pancreas makes extra insulin to make up for it. But, over time it isn't able to keep up and can't make enough insulin to keep blood glucose at normal levels.

Long-Term Effects – Over time, high blood sugar can damage and cause problems with:

- Heart and blood vessels
- Kidneys
- Eyes
- Nerves, which can lead to trouble with digestion, the feeling in the feet, and sexual response
- Wound healing

Causes of Diabetes – Usually a combination of things can cause type 2 diabetes, including:

- Metabolic syndrome.
  - People with insulin resistance often have a group of conditions including high blood glucose, extra fat around the waist, high blood pressure, and high cholesterol and triglycerides.
- Too much glucose from the liver.
  - When blood sugar is low, the liver makes and sends out glucose. After the patient eats, the blood sugar goes up, and the liver will slow down and store its glucose for later. But some people's livers don't. They keep cranking out sugar.
- Bad communication between cells.

---

[31] **Diabetes:**
> http://www.diabetes.org/diabetes-basics/type-2/#sthash.DFaAPiLP.dpuf
> **AND** http://www.webmd.com/diabetes/type-2-diabetes-guide/type-2-diabetes
> 08/30/2016

- o Sometimes cells send the wrong signals or don't pick up messages correctly. When these problems affect how the cells make and use insulin or glucose, a chain reaction can lead to diabetes.
- Broken beta cells.
  - o If the cells that make the insulin send out the wrong amount of insulin at the wrong time, the blood sugar gets thrown off.
  - o High blood glucose can damage these cells, too.

Symptoms: The symptoms of type 2 diabetes can be so mild the patient doesn't notice them. In fact, about 8 million people who have it don't know it.

- Being very thirsty
- Peeing a lot *(When it's not "blocked")*
- Blurry vision
- Being irritable *(Add that to the "other" irritability factors)*
- Tingling or numbness in the hands or feet
- Feeling worn out *(Always)*
- Wounds that don't heal *(See the next Chapter)*
- Yeast infections that keep coming back

Risk Factors and Prevention: While certain things make getting diabetes more likely, they won't give the patient the disease.

- Age: 45 or older *(Diagnosed at age 45)*
- Family: A parent, sister, or brother with diabetes *(in my case: my mother had diabetes)*
- Heart and blood vessel disease *(Certainly – I had/have that condition)*
- High blood pressure, even if it's treated and under control *(Certainly – I had/have that condition)*
- Low HDL ("good") cholesterol *(Certainly – I had/have that condition)*
- High triglycerides *(Certainly – I had/have that condition)*
- Acanthosis nigricans, a skin condition with dark rashes around the neck or armpits *(Yep! Add that to my list.)*
- Stress *(Certainly – I had/have that condition by virtue of dealing with all these "issues.")*

Chapter 20: **Wounds That Don't Heal**

HOSPILIZATION REPORT 12/18/2008: EXTREMITIES: Right
lower extremity with 3+ edema. Left lower extremity with
2+ edema. Right lower extremity with positive erythema,
tenderness to palpation, some scaling of the skin. The
swelling extends from the knee up to the foot, and there
is also 2+ ankle edema.

Erythema multiforme[32, 33] is a skin condition of unknown cause; it is a type of erythema possibly mediated by deposition of immune complexes (mostly IgM-bound complexes) in the superficial microvasculature of the skin and oral mucous membrane that usually follows an infection or drug exposure. It is an uncommon disorder, with peak incidence in the second and third decades of life. The disorder has various forms or presentations, which its name reflects (multiforme, "multiform", from multi- + formis). Target lesions are a typical manifestation. Two types, one mild to moderate and one severe, are recognized (erythema multiforme minor and erythema multiforme major).

The condition varies from a mild, self-limited rash (E. multiforme minor) to a severe, life-threatening form known as erythema multiforme major (or erythema multiforme majus) that also involves mucous membranes.

Consensus classification: Erythema multiforme major – typical targets or raised, edematous papules distributed acrally with involvement of one or more mucous membranes; epidermal detachment involves less than 10% of total body surface area (TBSA)

The mild form usually presents with mildly itchy (but itching can be very severe), pink-red blotches, symmetrically arranged and starting on the extremities. It often takes on the classical "target lesion" appearance, with a pink-red ring around a pale center. Resolution within 7–10 days is the norm.

---

[32] https://en.wikipedia.org/wiki/Erythema_multiforme   08/31/2016
[33] http://www.nhs.uk/conditions/erythema-multiforme/Pages/Introduction.aspx
08/31/2016

Individuals with persistent (chronic) erythema multiforme will often have a lesion form at an injury site, e.g. a minor scratch or abrasion, within a week. Irritation or even pressure from clothing will cause the erythema sore to continue to expand along its margins for weeks or months, long after the original sore at the center heals. (*I had two [2] "nasty" lesions on one leg that took almost a year to heal.*)

Erythema multiforme major (severe) or Stevens-Johnson syndrome: This rare form of the disease is much more severe, and can be life-threatening. It's usually caused by a reaction to mycoplasma bacteria or certain medications.

The rash is made up of bigger spots, which may fuse to produce large red areas. They may blister to reveal raw, painful sores. The patient might also have:

- a fever and headache, and feel unwell
- bloodshot or dry eyes that may burn, itch and weep
- sensitivity to light and blurred vision
- itchy skin
- achy joints
- septicemia
- myocarditis
- hepatitis
- hematuria
- acute tubular necrosis

What triggers it: In around half of cases, no cause can be found and nothing appears to trigger the symptoms. Bacterial infections and certain medications are thought to be possible triggers.

Infection: Most cases of erythema multiforme are caused by a viral infection – usually the herpes simplex (cold sore) virus. The cold sore virus lies dormant and tends to be reactivated by certain triggers. This explains why the condition can flare up repeatedly.

The second most common trigger is a chest infection caused by mycoplasma bacteria.

Medication: Medication can sometimes trigger the more severe type of erythema multiforme. Possible medications include:

- antibiotics such as sulfonamides, tetracyclines, amoxicillin and ampicillin
- anticonvulsants (used to treat epilepsy)
- non-steroidal anti-inflammatory drugs (NSAIDs)

Experts believe that an erythema multiforme reaction involves damage to the tissues and blood vessels of the skin.

How is it treated: Treatment aims to control the illness that is causing erythema multiforme, prevent infection and manage the symptoms.

Symptoms can be treated with:
- antihistamines to control the itching
- moist compresses held to the skin
- antiviral tablets, if the cause is a herpes simplex infection
- painkillers for skin pain

If the patient has Stevens-Johnson syndrome, the patient may need to stay in hospital and be treated in an intensive care unit (ICU) or a burns unit. The patient may also need:
- strong painkillers for the raw areas
- a short course of corticosteroid tablets to control the inflammation – only on specialist advice
- antibiotics if septicemia is suspected

Possible complications: Mild cases of erythema multiforme do not usually cause complications, and often gets better in two to three weeks using simple lotions. However, the disease can return, usually when the cause is the herpes simplex virus. In cases of Stevens-Johnson syndrome, possible complications can include:
- blood poisoning (septicemia)
- loss of body fluids and septic shock (where blood pressure drops to a dangerously low level)
- permanent skin damage and scarring
- skin infection (cellulitis)
- permanent eye damage

Occasionally, internal organs may be affected, causing inflammation of the heart (myocarditis), lungs (pneumonia), kidneys (nephritis) or liver (hepatitis).

More severe cases of Stevens-Johnson syndrome may take up to six weeks or longer to get better. A few people are left with scars on their skin after the rash clears up, and damaged vision if their eyes have been affected.

My legs are permanently diseased and disfigured – especially "below the knees." The open sores I acquired during the initial onset of the disease took 2 years to heal and left scarring. Even today, the skin on my legs will erupt (for no apparent reason) in oozing, bloody sores that may heal in a few weeks or linger for months.

My legs are dis-colored ranging from red to purple. The skin below my knees constantly flakes-off or peels in strips. My ankles and feet are covered in rough-dry skin that looks like it needs to be de-foliated.

I developed severe, painful, contact dermatitis that made my skin inflamed and sore. The disease flares up when my skin touches something it doesn't like (hence the name). It also made my skin incredibly dry and prone to cracking, which puts me at risk for secondary skin infections, which are sometimes worse than the condition itself.

During my 7-day hospital stay, a well-meaning dermatologist suggested the application of an oil-based cream to my legs to "retain the moisture." Needless to say my legs became angry red, swelled-up like balloons and the pain was excruciating! After washing the "moisturizer" off, my legs were able to breathe and got back to their normal size. No creams for me! I don't want to go there again! I will live with ugly legs…

Chapter 21: **I've Got a Lot of Nerve (Problems)**

But First – a word from our sponsor: Essential Tremor:[34]

Essential tremor is a nervous system (neurological) disorder that causes involuntary and rhythmic shaking. It can affect almost any part of the body, but the trembling occurs most often in the hands – especially when the patient does simple tasks, such as drinking from a glass or tying shoelaces.

It's usually not a dangerous condition, but essential tremor typically worsens over time and can be severe in some people. Other conditions don't cause essential tremor, although it's sometimes confused with Parkinson's disease.

Essential tremor signs and symptoms:
- Begin gradually, usually on one side of the body
- Worsen with movement
- Usually occur in the hands first, affecting one hand or both hands
- Can include a "yes-yes" or "no-no" motion of the head
- May be aggravated by emotional stress, fatigue, caffeine or temperature extremes

Essential tremor vs. Parkinson's disease: Many people associate tremors with Parkinson's disease, but the two conditions differ in key ways
- Timing of tremors:
    - Essential tremor of the hands usually occurs when the patient uses their hands.
    - Tremors from Parkinson's disease are most prominent when the hands are at your sides or resting in the lap.
- Associated conditions:
    - Essential tremor doesn't cause other health problems.
    - Parkinson's disease is associated with stooped posture, slow movement and shuffling gait.

---

[34] **Essential Tremor:**
   http://www.mayoclinic.org/diseases-conditions/essential-tremor/home/ovc-
   20177826   08/31/2016

- o However, people with essential tremor sometimes develop other neurological signs and symptoms, such as an unsteady gait (ataxia).
- Parts of body affected:
  - o Essential tremor mainly involves hands, head and voice.
  - o Parkinson's disease tremors usually start in the hands, and can affect legs, chin and other parts of the body.

Causes: About half of essential tremor cases appear to result from a genetic mutation, although a specific gene hasn't been identified. This form is referred to as familial tremor. It isn't clear what causes essential tremor in people without a known genetic mutation.

Risk factors:
- Autosomal dominant inheritance pattern
- Autosomal dominant inheritance pattern

Known risk factors for essential tremor include:
- Genetic mutation. The inherited variety of essential tremor (familial tremor) is an autosomal dominant disorder. A defective gene from just one parent is needed to pass on the condition.
- If the patient has a parent with a genetic mutation for essential tremor, the patient has a 50 percent chance of developing the disorder.
- Age. Essential tremor is more common in people age 40 and older.

Complications: Essential tremor isn't life-threatening, but symptoms often worsen over time. If the tremors become severe, the patient might find it difficult to:
- Hold a cup or glass without spilling
- Eat normally
- Put on makeup or shave
- Talk, if the voice box or tongue is affected
- Write legibly

Treatment: Some people with essential tremor don't require treatment if their symptoms are mild. But if the essential tremor is making it difficult to work or perform daily activities, discuss treatment options with the doctor.

Medications:

- Beta blockers. Normally used to treat high blood pressure, beta blockers such as propranolol (Inderal) help relieve tremors in some people. Side effects may include fatigue, lightheadedness or heart problems.

- Anti-seizure medications. Epilepsy drugs, such as primidone (Mysoline), may be effective in people who don't respond to beta blockers. Other medications that might be prescribed include gabapentin (Gralise, Neurontin) and topiramate (Topamax, Qudexy XR). Side effects include drowsiness and nausea, which usually disappear within a short time.

- Tranquilizers. Doctors may use drugs such as alprazolam (Xanax) and clonazepam (Klonopin) to treat people for whom tension or anxiety worsens tremors. Side effects can include fatigue or mild sedation. These medications should be used with caution because they can be habit-forming.

Therapy: Doctors might suggest physical or occupational therapy.

- Physical therapists can teach the patient exercises to improve muscle strength, control and coordination.

- Occupational therapists can help the patient adapt to living with essential tremor.

- Therapists might suggest adaptive devices to reduce the effect of tremors on daily activities, including:
  - o Heavier glasses and utensils
  - o Wrist weights
  - o Wider, heavier writing tools, such as wide-grip pens

I have successfully learned to cope with this ailment by "asking for help" in carrying things and using "special" (unorthodox) eating utensils – like big spoons so I can't spill stuff…

I (also) hold my hands still when I sit by either gripping something or by tucking them between the chair and my legs. Most people hardly notice anything. I am not "most people", my tremors can be pronounced, especially when I am trying to use my hands for tasks like changing small batteries, or using a screwdriver. Now, on to the "Main Character…"

DIABETIC NEUROPATHY:[35]

---

[35] **Diabetic Neuropathy:** http://www.mayoclinic.org/diseases-conditions/diabetic-neuropathy/basics/symptoms/con-20033336   08/31/2016

Symptoms: There are four main types of diabetic Neuropathy. You may have just one type or symptoms of several types.

*Definitely!! **I have all four of the below**!*

Most develop gradually, and the patient may not notice problems until considerable damage has occurred. The signs and symptoms of diabetic Neuropathy vary, depending on the type of Neuropathy and which nerves are affected.

Peripheral Neuropathy: the most common form of diabetic Neuropathy. The feet and legs are often affected first, followed by the hands and arms. Signs and symptoms of peripheral Neuropathy are often worse at night, and may include: Numbness or reduced ability to feel pain or temperature changes

- A tingling or burning sensation
- Sharp pains or cramps
- Increased sensitivity to touch – for some people, even the weight of a bed sheet can be agonizing
- Muscle weakness
- Loss of reflexes, especially in the ankle
- Loss of balance and coordination
- Serious foot problems, such as ulcers, infections, deformities, and bone and joint pain

Autonomic Neuropathy: The autonomic nervous system controls the heart, bladder, lungs, stomach, intestines, sex organs and eyes. Diabetes can affect the nerves in any of these areas, possibly causing:

- A lack of awareness that blood sugar levels are low (hypoglycemia unawareness)
- Bladder problems, including urinary tract infections or urinary retention or incontinence
- Constipation, uncontrolled diarrhea or a combination of the two
- Slow stomach emptying (gastroparesis), leading to nausea, vomiting, bloating and loss of appetite
- Difficulty swallowing
- Erectile dysfunction in men
- Increased or decreased sweating

- Inability of the body to adjust blood pressure and heart rate, leading to sharp drops in blood pressure after sitting or standing that may cause the patient to faint or feel lightheaded
- Problems regulating the body temperature
- Changes in the way the eyes adjust from light to dark
- Increased heart rate when the patient is at rest

Radiculoplexus Neuropathy (diabetic amyotrophy): affects nerves in the thighs, hips, buttocks or legs. Also called diabetic amyotrophy, femoral Neuropathy or proximal Neuropathy, this condition is more common in people with type 2 diabetes and older adults.

Symptoms are usually on one side of the body, though in some cases symptoms may spread to the other side. Most people improve at least partially over time, though symptoms may worsen before they get better. This condition is often marked by:

- Sudden, severe pain in the hip and thigh or buttock
- Eventual weak and atrophied thigh muscles
- Difficulty rising from a sitting position
- Abdominal swelling, if the abdomen is affected

Mononeuropathy: involves damage to a specific nerve. The nerve may be in the face, torso or leg. Mononeuropathy, also called focal Neuropathy, often comes on suddenly. It's most common in older adults.

Although mononeuropathy can cause severe pain, it usually doesn't cause any long-term problems. Symptoms usually diminish and disappear on their own over a few weeks or months. Signs and symptoms depend on which nerve is involved and may include:

- Difficulty focusing the eyes, double vision or aching behind one eye
- Paralysis on one side of the face (Bell's palsy)
- Pain in the shin or foot
- Pain in the lower back or pelvis
- Pain in the front of the thigh
- Pain in the chest or abdomen

Seek medical care if the patient notice:

- A cut or sore on the foot that doesn't seem to be healing, is infected or is getting worse

- Burning, tingling, weakness or pain in the hands or feet that interferes with the daily routine or sleep
- Dizziness
- Changes in digestion, urination or sexual function

These signs and symptoms don't always indicate nerve damage, but they may signal other problems that require medical care. Early diagnosis and treatment offer the best chance for controlling symptoms and preventing more-severe problems.

Even minor sores on the feet that don't heal can turn into ulcers. In the most severe cases, untreated foot ulcers may become gangrenous – a condition in which the tissue dies – and requires surgery or even amputation of the foot. Early treatment can help prevent this from happening.

Causes – Damage to nerves and blood vessels: Prolonged exposure to high blood sugar can damage delicate nerve fibers, causing diabetic Neuropathy. Why this happens isn't completely clear, but a combination of factors likely plays a role, including the complex interaction between nerves and blood vessels.

High blood sugar interferes with the ability of the nerves to transmit signals. It also weakens the walls of the small blood vessels (capillaries) that supply the nerves with oxygen and nutrients.

Other factors that may contribute to diabetic Neuropathy include:

- Inflammation in the nerves caused by an autoimmune response. This occurs when the immune system mistakenly attacks part of the body as if it were a foreign organism.
- Genetic factors unrelated to diabetes that make some people more susceptible to nerve damage.
- Smoking and alcohol abuse, which damage both nerves and blood vessels and significantly increase the risk of infections.

Risk factors: Anyone who has diabetes can develop Neuropathy, but these factors make the patient more susceptible to nerve damage:

- Poor blood sugar control. This is the greatest risk factor for every complication of diabetes, including nerve damage. Keeping blood sugar consistently within the target range is the best way to protect the health of the nerves and blood vessels.

- Length of time the patient has diabetes. Your risk of diabetic Neuropathy increases the longer the patient has diabetes, especially if blood sugar isn't well-controlled.
- Kidney disease. Diabetes can cause damage to the kidneys, which may increase the toxins in the blood and contribute to nerve damage.
- Being overweight. Having a body mass index greater than 24 may increase the risk of developing diabetic Neuropathy.
- Smoking. Smoking narrows and hardens the arteries, reducing blood flow to the legs and feet. This makes it more difficult for wounds to heal and damages the integrity of the peripheral nerves.

Diabetic Neuropathy can cause a number of serious complications, including:
- Loss of a limb. Because nerve damage can cause a lack of feeling in the feet, cuts and sores may go unnoticed and eventually become severely infected or ulcerated – a condition in which the skin and soft tissues break down.
  - The risk of infection is high because diabetes reduces blood flow to the feet.
  - Infections that spread to the bone and cause tissue death (gangrene) may be impossible to treat and require amputation of a toe, foot or even the lower leg.
- Charcot joint. This occurs when a joint, usually in the foot, deteriorates because of nerve damage.
  - Charcot joint is marked by loss of sensation, as well as swelling, instability and sometimes deformity in the joint itself. Early treatment can promote healing and prevent further damage.
- Urinary tract infections and urinary incontinence. Damage to the nerves that control the bladder can prevent it from emptying completely. This allows bacteria to multiply in the bladder and kidneys, leading to urinary tract infections. Nerve damage can also affect the ability to feel when the patient needs to urinate or to control the muscles that release urine.
- Hypoglycemia unawareness. Normally, when the blood sugar drops too low – below 70 milligrams per deciliter (mg/dL), or 3.9 millimoles per liter (mmol/L) – the patient develops symptoms such as shakiness, sweating and a fast heartbeat.

Autonomic Neuropathy can interfere with the ability to notice these symptoms.

- Low blood pressure. Damage to the nerves that control circulation can affect the body's ability to adjust blood pressure. This can cause a sharp drop in pressure when the patient stands after sitting (orthostatic hypotension), which may lead to dizziness and fainting.

- Digestive problems. Nerve damage in the digestive system can cause constipation or diarrhea – or alternating bouts of constipation and diarrhea – as well as nausea, vomiting, bloating and loss of appetite. It can also cause gastroparesis, a condition in which the stomach empties too slowly or not at all. This can interfere with digestion and cause nausea, vomiting and bloating, and severely affect blood sugar levels and nutrition.

- Sexual dysfunction. Autonomic Neuropathy often damages the nerves that affect the sex organs, leading to erectile dysfunction in men and problems with lubrication and arousal in women.

- Increased or decreased sweating. When the sweat glands don't function normally, the body isn't able to regulate its temperature properly. A reduced or complete lack of perspiration (anhidrosis) can be life-threatening. Autonomic Neuropathy may also cause excessive sweating, particularly at night or while eating.

I have learned to "cope" with the pain of diabetic neuropathy. In my case, the symptoms of diabetic neuropathy are pain and numbness in my extremities, problems with my digestive system, urinary tract, blood vessels and heart. My diabetic neuropathy is quite painful almost to the point of being disabling. I have been told it can be fatal.

Just when it seems I "have it under control" and I have not experienced a debilitating attack for several dyas – the pain returns as a stabbing, knifing sensation. My legs, ankles and feet are the worst pain.

Chapter 22: **Gout & About**

Gout[36] is characterized by sudden, severe attacks of pain, redness and tenderness in joints, often the joint at the base of the big toe.

Gout – a complex form of arthritis – can affect anyone. Men are more likely to get gout, but women become increasingly susceptible to gout after menopause.

An attack of gout can occur suddenly, often waking the patient up in the middle of the night with the sensation that their big toe is on fire. The affected joint is hot, swollen and so tender that even the weight of the sheet on it may seem intolerable.

The signs and symptoms of gout almost always occur suddenly – often at night – and without warning. They include:

- Intense joint pain: Gout usually affects the large joint of the big toe, but it can occur in feet, ankles, knees, hands and wrists.
  - The pain is likely to be most severe within the first four to 12 hours after it begins.
- Lingering discomfort: After the most severe pain subsides, some joint discomfort may last from a few days to a few weeks.
  - Later attacks are likely to last longer and affect more joints.
- Inflammation and redness: The affected joint or joints become swollen, tender, warm and red.
- Limited range of motion: Decreased joint mobility may occur as gout progresses.

Gout occurs when urate crystals accumulate in the joint, causing the inflammation and intense pain of a gout attack. Urate crystals can form when the patient has high levels of uric acid in their blood.

The body produces uric acid when it breaks down purines – substances that are found naturally in your body, as well as in certain foods, such as steak, organ meats and seafood. Other foods also promote higher levels

---

[36] **Gout:**
http://www.mayoclinic.org/diseases-conditions/gout/basics/definition/con-20019400   09/03/2016

of uric acid, such as alcoholic beverages, especially beer, and drinks sweetened with fruit sugar (fructose).

Normally, uric acid dissolves in the blood and passes through the kidneys into your urine. But sometimes the body either produces too much uric acid or the kidneys excrete too little uric acid. When this happens, uric acid can build up, forming sharp, needle-like urate crystals in a joint or surrounding tissue that cause pain, inflammation and swelling.

A patient is more likely to develop gout if they have high levels of uric acid in their body. Factors that increase the uric acid level in the body include:

- Diet. Eating a diet that's high in meat and seafood and high in beverages sweetened with fruit sugar (fructose) promotes higher levels of uric acid, which increases the risk of gout.
    - Alcohol consumption, especially of beer, also increases the risk of gout.
- Medical conditions. Certain diseases and conditions make it more likely that gout will develop.
    - These include untreated high blood pressure and chronic conditions such as diabetes, metabolic syndrome, and heart and kidney diseases.
- Certain medications. The use of thiazide diuretics – commonly used to treat hypertension – and low-dose aspirin also can increase uric acid levels.
- Family history of gout. If other members of a family have had gout, a patient is more likely to develop the disease.
- Age and sex. Gout occurs more often in men, primarily because women tend to have lower uric acid levels. After menopause, however, women's uric acid levels approach those of men. Men also are more likely to develop gout earlier – usually between the ages of 30 and 50 – whereas women generally develop signs and symptoms after menopause.
- Recent surgery or trauma. Experiencing recent surgery or trauma has been associated with an increased risk of developing gout.

People with gout can develop more-severe conditions, such as:

- Recurrent gout. Some people may never experience gout signs and symptoms again. But others may experience gout several times each year.

- o Medications may help prevent gout attacks in people with recurrent gout.
- o If left untreated, gout can cause erosion and destruction of a joint.
- Advanced gout. Untreated gout may cause deposits of urate crystals to form under the skin in nodules called tophi (TOE-fie). Tophi can develop in several areas such as your fingers, hands, feet, elbows or Achilles tendons along the backs of the ankles.
  - o Tophi usually aren't painful, but they can become swollen and tender during gout attacks.
- Kidney stones. Urate crystals may collect in the urinary tract of people with gout, causing kidney stones.
  - o Medications can help reduce the risk of kidney stones.

I suffer from "Recurrent Gout" despite the best efforts of medication (Allopurinol)

EPILOG

Have I answered the question: "What's wrong with you?" I think so and the answer is; "There's a lot of things wrong with me."

Admittedly, some of the items can be "nuisance-diseases", but (actually) they add connections to the maddening list of: "What's wrong with you?"

During my research I discovered that Medical Staff, Insurance Companies, Hospitals and others have devised a method to help quantify illnesses and diseases in individuals. They use the term "Morbidity."

Morbidity (from Latin *morbidus,* meaning "sick, unhealthy")[37] is a diseased state, disability, or poor health due to any cause. The term may be used to refer to the existence of any form of disease, or to the degree that the health condition affects the patient. Among severely ill patients, the level of morbidity is often measured by various ICU* scoring systems.

* *An Intensive Care Unit (ICU), also known as an Intensive Therapy Unit or Intensive Treatment Unit (ITU) or Critical Care Unit (CCU), is a special department of a hospital or health care facility that provides intensive treatment medicine.*

In my case I am a "**Comorbid Patient**," since I have more than one active disease.

In medicine, **Comorbidity**[38] is: the presence of one or more additional diseases or disorders co-occurring with (that is, concomitant or concurrent with) a primary disease or disorder; in the countable sense of the term, a comorbidity (plural comorbidities) is each additional disorder or disease.

The Charlson comorbidity index[39] predicts the one-year mortality for a patient who may have a range of comorbid conditions, such as heart

---

[37] https://en.wikipedia.org/wiki/Disease#Terminology   09/01/2016
[38] https://en.wikipedia.org/wiki/Comorbidity#Diagnosis-related_group   09/01/2016
[39] https://en.wikipedia.org/wiki/Comorbidity#Charlson_index   09/01/2016

disease, AIDS, or cancer (a total of 22 conditions). Each condition is assigned a score of: 1, 2, 3, or 6, depending on the risk of dying associated with each one. Scores are summed to provide a total score to predict mortality. Clinical conditions and associated scores are as follows:

- 1 each: Myocardial infarct, congestive heart failure, peripheral vascular disease, dementia, cerebrovascular disease, chronic lung disease, connective tissue disease, ulcer, chronic liver disease, diabetes.
- 2 each: Hemiplegia, moderate or severe kidney disease, diabetes with end organ damage, tumor, leukemia, lymphoma.
- 3 each: Moderate or severe liver disease.
- 6 each: Malignant tumor, metastasis, AIDS.

My personal (calculated) Charlson Score = **12** [40] (Feel free to calculate your own number.)

Interpretation: **My 10 Year Survival Percentage = 0.00%** *(I go back in time to 1989 when I received my diagnosis of Focal Glomerular Sclerosis – the Kidney doctor was very somber in his estimation that without his "special" regimen of steroid treatments I would be dead in six years…I said "No thank you," and here we are, 27 years later. No one knows the hour, the day, the month or the year anyone else will die. ONLY when you are called by your creator – is it the final call.)*

It is interesting to note that the term "Multimorbidity" is sometimes used to describe several medical diseases in the same individual and is interchangeable with Comorbidity. **Multimorbidity**[41] is defined as the co-occurrence of two or more chronic medical conditions in one person. Unfortunately, some medical organizations and professionals tend to equate this to a "survivability rate," or to put things more crassly: *"This person isn't going to be alive in ten years and we (I) can't change things, no matter what we do."*

Of course, I beg to differ.

---

[40] **Charlson-Comorbidity-Index-(CCI)-Calculator**:
http://www.thecalculator.co/health/Charlson-Comorbidity-Index-(CCI)-Calculator-765.html   09/02/2016
[41] https://en.wikipedia.org/wiki/Disease#Terminology   09/01/2016

I have never sought a "cure" – that would be an unrealistic expectation. I have always gravitated towards stabilization of the disease progression (whenever possible) and relief from debilitating symptoms. Since I AM "still alive" I feel I have been successful in achieving this goal.

Could my "secret" be one of the following?

- Consuming the "Mushroom of Immortality", a key ingredient in the elixir of life. The Lingzhi mushroom, literally translated as the "Supernatural Mushroom," is the oldest known mushroom used medicinally.

- Daily consumption of "Amrita?" According to the Rigveda, (a collection of ancient Vedic hymns that are a cornerstone of Hinduism), Amrita is a drink that bestows immortality. In Hinduism and other traditions, it is also referred to as Soma. Indra, (the god of heaven) and Agni, (the god of fire,) drink Amrita to attain immortality. After drinking the mysterious substance, they state: *"We have drunk Soma and become immortal; we have attained the light, the Gods discovered. Now what may foeman's malice do to harm us? What, O Immortal, mortal man's deception?"* (Rigveda 8.48.3)

- OR did I move to Pennsylvania to partake of the Philosopher's Stone? Legend claims that the Philosopher's Stone [42] is in a creek in Philadelphia. A 17th century group called the "Society of the Woman in the Wilderness" settled in the woods outside of Philadelphia's Germantown section. The group was led by German pietist and occultist Johannes Kelpius, who believed the world would end in 1694. (It did not.) The group spent much of its time in peaceful meditation in caves and modest homes on the outskirts of the city. After Kelpius's death, some of his students claimed that he had been the guardian of the Philosopher's Stone, which he kept hidden in his meditation cave. Immediately before his death, it is said he ordered his students to toss the stone into the nearby Wissahickon Creek. The cave is still accessible, and is marked as a historic site today.

---

[42] The philosopher's stone, or stone of the philosophers (Latin: lapis philosophorum) is a legendary alchemical substance capable of turning base metals such as mercury into gold (chrysopoeia, from the Greek χρυσός khrusos, "gold," and ποιεῖν poiein, "to make") or silver. It is also able to extend one's life and called the elixir of life, useful for rejuvenation and for achieving immortality; for many centuries

There is no truth to any of those suppositions. I credit my daily "survival" first to the Will of God and second to a rational approach to "knowing my limits."

Here are some important points:

- I don't get "angry" at my diseases. Anger has never proven to be a cure for anything. I have been blessed to be an example to others who wish to see me and understand me.

- I DO get frustrated and upset when my diseases and their symptoms seem to overwhelm my everyday life. I actually do not enjoy (nor do I want to enjoy) experiencing daily diarrhea, stomach cramps, shortness of breath, painful bouts of Neuropathy, dyspnea, etc., etc. I am not a martyr!

- My frustration usually "boils over" to a nasty personality which sometimes is misinterpreted as anger towards individuals or anti-social behavior. Yes, I am wrong to project those emotions, but that's the "sinner in me"

- I also realize that the simplest things are quickly eluding me. My hands shake, I forget things, I can't stabilize myself, loud noises make me "erupt." As before, more frustration…

- I don't sit and "fester" about how sick I am. There are times however, when I face the realization that I may be "re-called" at any moment and my sadness is knowing the loss of my loved ones and the mental suffering they will have to endure. I don't want them to care, to feel any sadness because of me, but I fear they will.

- I don't believe in "whacko" cures or regimens:
  - Vegetable juicing, immune-boosting supplements, stress reducing techniques.
  - Leeches
  - Coffee Enemas
  - Energy-Deflecting Golfer Pendants
  - The Dr. Budwig Protocol – consuming a mixture of cottage cheese, flaxseeds, and flaxseed oil
  - Vitamin C Chelation
  - Frankincense Essential Oil Therapy (another Dr. Budwig idea)
  - Turmeric and Curcumin
  - Immune-Boosting Mushrooms
  - Apricot kernels

Some things that HAVE "worked" for me"
- Probiotic Foods & Supplements – especially after colon procedures.
- Vitamin D3 – I have a proven vitamin D deficiency.
- Prayer and Building Peace.

The "discussion part" is now over. Now, the "Tough-Love" part begins.

Every day, I start my day with a "clean-slate," determined not to let any of my diseases or symptoms ruin my life. Then comes the pain, from somewhere, or the breathing difficulties, or the dizziness…the pee or the poo, it doesn't matter, it happens. I try my best to hide my frustration of not being able to lead a "normal life" (sometimes, I can barely "dust" my bedroom furniture without breaking something or almost falling down). Then, what I call the "brain fog" sets in, as the toxins build-up and my kidneys and liver fail to do their jobs. I become physically exhausted just by staying awake – by 5 PM I am a "physical wreck." The misunderstandings begin:

"Do you have to act that way?" Answer: No, I don't, but I just can't "switch-off" each pain or symptom by waving my magic wand. Very often if I get rid of my breathing problem (as an example), some other unusual pain, weakness or "constitutional failing" will take its place. (Maybe a bloody nose…)

"Let me help you." Answer: I appreciate your help, but please realize that I feel more frustration (or helplessness) in knowing that I can't even pour myself a glass of milk.

"You need to do (whatever)." Answer: I need to cope with the situation in a way that helps me get out of my downward spiral and that my way may not be the "standard (or accepted) way."

I would appreciate a question, now and then, such as: "How are you doing today?" Asked with the understanding that I may NOT be doing well, but I'm trying to "act well."

Another great question might be: "Do you need anything?" This would be helpful when I am sitting "stock-still" in my recliner trying to endure the latest onslaught of: (substitute any discomfort here).

It's a "good day" for me if I don't disrespect too many people and can cope with the discomfort. All I ask from anyone is forgiveness of my "bad attitude" and a little understanding.

Now, if you'll excuse me, I feel some stomach cramps coming on…

One Last Thing

I want to provide some insight concerning my hospital stays and how doctors, nurses and staff can better meet my needs:

1. Be careful with the blood thinners! I realize there are "protocols," however I tend to bleed profusely when given blood thinners. I bleed from my nose (mostly), although this will progress to my crotch in the form of a bloody rash and then my armpits.
2. Have a bed available that I can "get into" and is not sized for a circus gymnast who can contort their body into a figure eight. I am NOT physically "flexible". I cannot climb, jump, "skooch" or otherwise position myself. I require a large bed that can "kneel" to accommodate my disabilities.
3. The room should be no warmer than 70^0 in the daytime and 65^0 at night. I will become overheated and "grouchy."
4. If you ask me a question and I give you an answer, don't tell me: "That's not the right answer." I can tell you many things about my body that you do not know. I have had many "adventures" in medicine that were not covered in the classroom – I know my body. My body does not fit your standard protocols, consider me an "outlier".[43]
5. No "creams" or "moisturizers", ever! The quickest way to ensure I get "sicker" is to stop my skin from breathing by slathering creams all over me.
6. My bathroom visits are numerous and follow an erratic schedule.
7. Phlebotomist listen up! My skin is as "tough" as an elephant's and you may think you have found a good vein…BUT…touch the needle to my skin and "away it goes." Try again, the current world record for one arm is five tries.
8. I cannot breathe and control my heart-rate/blood pressure. If you ask me to move around and then take my blood pressure it will be "high." If I am gasping for breath, do not expect my "numbers' to be "normal." Also, use a "wide-cuff" for blood pressure, please! I can guarantee the readings will be wrong if you use a "granny cuff."
9. If my IV is "leaking," please have the courtesy to start a new line instead of handing me a box of tissues to soak-up the medicine.
10. I have a nerve disorder called "Cutaneous Allodynia." Normally it's no big deal, except that my "hair hurts" if it is too long, or my fingernails bother me if they are not trimmed. Sometimes, my clothes bother me. Just be aware that bandages and

---

[43] Something that lies outside the main body or group that it is a part of. In Math, a value that "lies outside" (is much smaller or larger than) most of the other values in a set of data.

"wrappings" are not my friends – I literally want to rip them off after a few minutes! Fighting these urges makes me "grumpy."

11. I really don't care about your trip to Cancun!

12. I know I often have either a "low" body temperature (96 +/-) or a FUO (fever – unknown origin). Again, I am not your standard patient.

13. My oxygenation level (without O_2 supplement) will be < 94%. I will struggle to catch my breath most times.

14. I suffer bouts of Dyspnea – the sensation of difficult or uncomfortable breathing.

15. More often I experience Orthopnea – the sensation of breathlessness in the recumbent position, relieved by sitting or standing.

16. I (also) Paroxysmal nocturnal dyspnea (PND) – a sensation of shortness of breath that awakens me, often after 1 or 2 hours of sleep, and is relieved in the upright position. (#14, 15 & 16 are a result of my heart conditions)

17. Approach me as a "System." Individual doctors and medical protocols tend to focus on "specialty" treatments. For example: A doctor might assume I am "retaining fluids" and prescribe "water-pills/medications" – to make me urinate more frequently; whereas a Nephrologist might see that treatment as injurious to my renal function and can increase blood glucose level (which would upset the Endocrinologist). Consider what the "systemic effects" of the treatment have on me as a patient, given my overall history and my current condition.

18. Try not to "talk down" to me. I will be a more "Happy Camper" if you use the real terms. Instead of saying: "We are going to check your heart," say: "You are going to have an Echocardiogram." Chances are, I've had the test before…sometimes several times.

19. I often remember the "Code of Conduct" from my military days, "…I will make every effort to escape…" I want to leave the hospital as soon as I can. I have no intentions of staying "another night." My fastest recuperations have been at home and I intend to stress that to the hospital staff.

– The End –

APPENDIX: Includes all available scans/reports

NOTES:
1. All personal, or other identifying information, has been redacted from the reports.
2. The actual format of each report is presented whenever possible, extraneous information, watermarks and trademarks have been deleted.
3. Aortic Valve and Heart Studies reports are included in the Chapter "My Heart" and not repeated here.

#	Report Content	Date	Page
1	Kidney Biopsy Report	01/03/1989	88
2	Colon Polyp: Pathology Report	06/23/1994	90
3	Abdominal Ultrasound	02/18/2006	91
4	CT Scan – Abdomen/Pelvis	03/07/2006	93
5	CT Scan – Full Torso	11/30/2012	94
6	Ultrasound – Carotid Arteries	06/25/2014	97
7	Ultrasound – Retroperitoneal Comp (Kidney & Bladder)	01/19/2015	98
8	Ultrasound – Abdomen	03/27/2015	99
9	CT virtual colonoscopy	09/18/2015	100
10	Prescription Drug List	Current	102

1. Kidney Biopsy Report

NEW ENGLAND DEACONESS HOSPITAL
WILLIAM A. MEISSNER LABORATORY OF PATHOLOGY
185 PILGRIM ROAD, BOSTON, MASSACHUSETTS 02215

Shields Warren, M.D. Founder   William A. Meissner, M.D. Emeritus
Merle A. Legg. M.D. Chairman

Geung H. Ahn, M.D.	Paola C. DeGirolami, M.D.	William T. Lockard. Jr. M.D.
Hans T. Aretz. M.D.	Walter H. Dzik, M.D.	Takashi Maki, M.D. Ph.D.
Charles F. Arkin, M.D.	Micheline Federman. Ph.D.	Carl J. O'Hara. M.D.
Karolyn Balogh, M.D. M.D.	Urmila Khetlry, M.D.	Joseph J. Schildkraut,
Ronald G. Bardawil, M.D.	Agnes Kim, M.D.	Mark L. Silverman. M.D.
Ann W. Crosson. M.D.	Arthur K. Lee, M.D.	

BISOL, JOHN L. Patient: 40407918 – 01/03/1989
Dr. Gomery – ADMITTING
Dr. Rolla – SURGEON
Dr. Diamond – KIDNEY

Specimens:
  A. Specimen A consists of a tan brown core measuring 1.3 cm.in length and up to 0.1 cm. in width.
  B. Specimen B consists of two tan brown wedge shaped fragments measuring 0.8 x 0.4 x 0.2 cm. and 0.7 x 0.4 x 0.2 cm., respectively.

Pathologic Diagnosis:
  A. Unremarkable medulla. No glomeruli found.
  B. Focal global and segmental glomerulosclerosis, probably secondary focal glomerulosclerosis. Hyaline arteriolosclerosis.

Light microscopy: Specimens consist of an open renal biopsy containing numerous glomeruli, the vast majority of which appear enlarged with normal to focally increased mesangium. Rare glomeruli show segmental sclerosis, some with focal hyalinosis. Somewhat more show global sclerosis, sometimes grouped with interstitial fibrosis and tubular atrophy, forming a microscar Other areas show small foci of tubular atrophy and interstitial fibrosis. Focal hyaline arteriolosclerosis is noted.

Immunofluorescence (Performed by Dr. H. Rennke, Brigham & Women's Hospital: The sections. were reacted with antibodies against IgG, IgA, IgM, C3, albumin, fibrinogen, kappa and lambda light chains There are

approximately 30 glomeruli in the sample. Three
glomeruli show segmental coarse staining for C3 (2-3+,
IgM (2+) and fibrin (1+) Some tubular cst9 react
strongly with kappa light chain and IgA.

Electron microscopy (Performed by Dr. M. Federman): Two
glomeruli are examined. The first shows normal to
minimally thickened basement membranes with only focal
effacement of epithelial foot processes. Mesangium
focally is increased. No electron dense deposits are
found. A few areas show apparent mesangial interposition in the basement membrane. A few flecks of
fibrin are present in the lumen. The second glomerullus
shows extensive alteration in capillary loops and
basement membranes with more extensive apparent
mesangial interposition and sub endothelial changes,
probably reflecting endothelial damage. Electron dense
material, consistent with hyalinosis, is present
focally. There is focal effacement of epithelial foot
processes.

Note: This probably represents a secondary focal
glomerulosclerosis because of the. enlarged glomeruli,
only focal effacement of the epithelial foot processes
by EM and the clinical history rather than primary
glomerulosclerosis.

Reference: Kasiske, Bertram and Crosson, John T. Renal
Disease in Patients

2. Colon Polyp: Pathology Report

MARLBOROUGH HOSPITAL

A member of UMass Memorial Health Care

P a t h o l o g y : R e p o r t

Patient: BISOL, JOHN L
DOB: Age:
Sex: M

Consultation Requester: ERNESTO Y. JOSE, M.D.

Laboratory No.: S-94-1464   Hospital No.: 643199
Date Received: 06/23/94   MedRec No.: 097653
Pre-OP Diagnosis: POLYP
Post-Op Diagnosis:

GROSS DESCRIPTION:

SPECIMEN LABELED SIGMOID CONSISTS OF A SPHERICAL,
RUBBERY, BLUISH-RED, SMOOTH NODULE MEASURING 2 X 1.8 X
1.8 CM. SECTIONING REVEALS HOMOGENEOUS YELLOW TISSUE
COVERED BY FOCALLY ULCERATED MUCOSA.
RSS
LC: FF
MICROSCOPIC DIAGNOSIS:
COLON:
      SUBMUCOSAL LIPOMA, PARTIALLY INFARCTED.

LC: PL

                              LUIGI CECCACCI M.D.
                              Pathologist

---

3. Abdominal Ultrasound

MARLBOROUGH HOSPITAL

A member of UMass Memorial Health Care

R a d i o l o g y

Patient: BISOL, JOHN L
DOB: **Age:**
Sex: M

Consultation Requester: ERNESTO Y. JOSE, M.D.

MRN: MM0097653   Location: MM-US Status: OUT-PATIENT
Tech: NOEL, DZEJA Visit number: MM000671300

HISTORY: ABD US (ATT LIVER/BILIARY TREE) - RUQ PAIN US,
ABDOMEN, B_MODE JOSE, ERNESTO Y 18-Feb-2006 12:36 PM
      6732134

Final Report - ABDOMINAL ULTRASOUND

HISTORY: Right upper quadrant pain. The liver is
extensively echogenic and cannot be completely
penetrated. The pancreas and liver are not seen a
result. There is no obvious intrahepatic biliary ductal
dilatation. The common duct is enlarged to 11mm. The
gallbladder has been removed.

Right kidney measures 12.2cm in length and the left
14.1cm in length. There are several cysts present. In
the right upper pole there is 3cm x 3.2cm cyst and in
the lower pole 2.7 x 2.8cm cyst. In the upper pole of
the left kidney, there is a 1.5cm x 2.8cm simple cyst.
In the lower to mid pole, there is a 3.8 x 3.4cm
septated cyst. The lower portion of the septation
appears to have another septation within it and it
appears complex in nature.
Follow-up examination with a CT examination is
recommended.

The spleen is enlarged at greater than 15cm. Normal
direction portal venous flow is demonstrated with color
Doppler imaging. No free fluid is seen.

IMPRESSION:

- o Extensively echogenic liver and enlarged spleen suggesting cirrhosis or extensive fatty changes, however there is no free fluid and there is normal direction portal venous flow.
- o The pancreas and aorta cannot be visualized.
- o There is enlargement of the extra hepatic biliary ducts post cholecystectomy.
- o Questionable complex cystic focus within a bilocular cyst in the left mid to lower kidney for which additional CT scan is recommended.
- o In addition, CT examination can visualize the aorta and pancreas, which were not seen on this examination.

4. CT Scan – Abdomen/Pelvis

MARLBOROUGH HOSPITAL

A member of UMass Memorial Health Care

R a d i o l o g y: C o n s u l t a t i o n

Requester: ERNESTO Y. JOSE, M.D.
Status: OUTPATIENT   **Tech:** ZONA, DAVID J.
Visit number: 7-Mar-2006 4:31 PM 6750818
Patient: John L. Bisol   **Age:**
History: CT ABD - RUQ PAIN

F i n a l R e p o r t

CLINICAL HISTORY: 57-year-old male with right upper quadrant pain.

TECHNIQUE: Contiguous 5 mm. axial images were obtained from the level of the lung bases to the level; of the pubic symphysis after the administration of oral contrast. Intravenous contrast was not administered per request of the referring physician. Coronal reformatted images were obtained.

FINDINGS: The lung bases are clear bilaterally.
The liver and spleen are homogeneous in attenuation and are normal in size.
The patient is status post cholecystectomy.
The stomach, pancreas and adrenal glands are normal.
There are bilateral renal cysts. A 4-cm. cyst is present laterally within the left kidney. A 3.2-cm. cyst is present within the upper pole of the right kidney. A 2.5-cm. cyst is present within the lower pole of the right kidney.
There is no evidence of hydronephrosis bilaterally.
Diverticula are present arising from the descending and sigmoid colon.
There is no evidence of diverticulitis. There is no evidence of adenopathy within the abdomen or pelvis. The abdominal aorta is unremarkable. The urinary bladder and prostate gland are normal. There are no bony abnormalities.

IMPRESSION:
o Status post cholecystectomy.
o Bilateral renal cysts.
o Diverticulosis with no evidence of diverticulitis.

5. CT Scan – Full Torso

Abington Health
Lansdale Hospital
Department of Radiology – Radiology Group of Abington, PC

Patient Name: BISOL, JOHN
DOB: **Sex:**   M   **Status:** O
Pt. Location: LREG

Exam: CT CAP WO CONTRAST   **ACC:** 8552958 (LH) CTCAP
CPT: 71250
Date/Time Completed: 11/30/2012 9:36:00AM
Signs & Symptoms: 786.6 Chest swelling/mass/lump
Requesting Provider:
BOTHWELL, WILLIAM N

HISTORY: Left chest and abdominal wall swelling and soreness, chronic renal insufficiency

COMMENT: Following oral contrast but without intravenous contrast, axial, coronal, and sagittal images were obtained through the chest, abdomen, and pelvis using standard protocol on a 64-MDCT.

COMPARISON: None

Study was ordered without intravenous contrast because of elevated creatinine.

The study is limited by large patient body habitus.

The lower thorax and anterolateral abdominal wall extend beyond the field of view and are suboptimally evaluated.

- There is mild asymmetric gynecomastia greater on the left.
- There are lower ventral abdominal wall skin thickening and subcutaneous fatty induration suggesting mild panniculitis. No abscess is seen.
- There is mild bibasilar linear atelectasis/scarring. Lungs are otherwise clear.
- There is moderate eventration of the right Hemidiaphragm.
- There is no pathologic-sized mediastinal, hilar, or axillary adenopathy on this unenhanced scan.
- There are tiny hypodense thyroid nodules bilaterally.
- Heart size is within normal limits.
- There are dense coronary artery and aortic valve calcifications.
- There is no pericardial effusion.
- The unenhanced liver is enlarged with a maximal craniocaudal length of 19.9

cm.
- The unenhanced spleen is mildly enlarged with a craniocaudal length of 13.2 cm.
- The gallbladder is surgically absent.
- There is a small mildly complex right renal upper pole cyst with wall calcification.
- There is a bilobed 6.7 x 3.7 x 3.8 cm exophytic right lower pole cyst.
- There is a 1.6 x 1.3 x 1.6 cm isodense complex cyst versus solid mass arising from the left upper pole.
- There is a small cyst in the left midpole.
- There is a small left lower pole parapelvic cyst.
- There are a few very small bilateral renal hypodensities, which are too small to characterize.
- There is no hydronephrosis or nephrolithiasis.
- The urinary bladder is incompletely distended.
- Prostate is mildly enlarged with punctate calcifications.
- The unenhanced pancreas and adrenal glands are grossly unremarkable.
- There is suboptimal oral opacification with no oral contrast in the distal small bowel or in the colon.
- There are a few mildly distended nonspecific small bowel loops in the right abdomen. (Obstruction is not suggested.)
- The colon is redundant and contains mild to moderate fecal material.
- There are multiple colonic diverticula.
- There is no significant mural thickening.
- The appendix is not visualized with certainty, but there are no secondary signs for acute appendicitis.
- There is mild contrast in the distal esophagus suggesting gastroesophageal reflux or stasis.
- There is a very small hiatal hernia.
- There is no significant free fluid or fluid collection.
- There are small fat-containing bilateral inguinal hernias.
- There is no pathologic-sized adenopathy in the abdomen and pelvis.
- There are mild to moderate vascular calcifications.
- There are degenerative changes in the spine, sacroiliac joints, and hips.

IMPRESSION:
- Limited study without significant chest wall inflammatory/infectious abscess or mass.
- Mild left greater-than right asymmetric gynecomastia.
- Mild lower ventral wall skin thickening and fatty induration suggests panniculitis.
- Tiny hypodense thyroid nodules.
- Hepatosplenomegaly
- Small left renal isodense complex cyst versus solid mass.

- ○ Moderate colonic diverticulosis.
- ○ Nonspecific but non-obstructive mildly distended right abdominal small bowel loops.

Signed by: PAN, JEFFREY Signed on: 11/30/2012 11:44:44

6. Ultrasound – Carotid Arteries

Abington Health
Lansdale Hospital

Department of Radiology – Radiology Group of Abington, PC

Requestor: Lansdale   TVPC Jerome E.Sag, M.D.
Patient: John Bisol
Ordered Date: 06/25/2014
Test Name: Carotid Imaging Bilateral

The study is technically limited.

The right common carotid, carotid bulb and the proximal internal and external carotid arteries are visualized. There is probable mild plaque formation in the right carotid bulb.

Doppler tones and color-flow are obtained in all vessels with peak velocity of 85/13 in the common carotid artery, 66/6 in the internal carotid artery and peak systolic velocity of 80 in the external carotid artery. The right vertebral artery flow is antegrade.

The left common carotid, carotid bulb and the proximal internal and external carotid arteries are visualized. There is no clear plaque formation seen. Doppler tones and color-flow are obtained in all vessels with peak velocity of 69/13 in the common carotid artery, 61/8 in the internal carotid artery and peak systolic velocity of 189 in the external carotid artery.

The left vertebral artery flow is antegrade.

Impression:
   o Probable mild stenosis at the level of the right carotid bulb.
   o Unremarkable left carotid study.

Signed by: WANG, CINDY
Signed on: 06/25/201418:07:47
Result: Received 06/25/2014

7. Ultrasound – Retroperitoneal Comp (Kidney & Bladder)

Abington Health
Lansdale Hospital
Department of Radiology – Radiology Group of Abington, PC

Requestor: Lansdale TVPC Prashant P. Parikh, M.D.
Patient: John Bisol
Ordered Date: 01/19/2015
Assessments: UTI
Test Name: US Retroperitoneal Comp (Kidney & Bladder)
Name Value: Ultrasound retroperitoneum.

History: Urinary tract infection, difficulty urinating, cholecystectomy

Comparison: CT November 30, 2012
Ultrasound examination of the kidneys and urinary bladder was performed.

The right kidney measures 12.4 x 6.1 x 4.8 cm.

The left kidney measures 12.3 x 5.1 x 6.4 cm.

There is a complex renal cyst at the upper pole with an internal septation, measuring 4.4 x 3.6 by 4.7 cm; previously measuring 3 x 3.8 X 3.6 centimeter in November 2012 CT.

Parapelvic cyst left midpole kidney measures 1.7 x 1.3 by 1.6 cm.

There is no hydronephrosis or calculus.

There is normal renal parenchymal echogenicity.

The urinary bladder prior to voiding measures 8.1 by 12.2 x 6.4 cm for a volume of 443 ml.

After voiding the urinary bladder measures 6.3 x 3.3 by 4.5 cm for a volume of 49 ml.

Impression:
- o   No hydronephrosis or calculus
- o   49 ml post void residual urine
- o   Complex cyst left kidney is slightly larger than 2012 study

Signed by: LIM, M.D., PHILIP S. 01/23/2015 09:00:34

8. Ultrasound – Abdomen

Abington Health
Lansdale Hospital
Department of Radiology – Radiology Group of Abington, PC

Patient: JOHN BISOL,   March 27, 2015, 15:30
Performer: DONALD ZAJICK,
Author: DONALD ZAJICK, Radiology Group of Abington
Encounter Date: at March 27, 2015, 14:52:00
Provider: DIVO MESSORI,
Information recipient: DIVO MESSORI,
Legal authenticator: DONALD ZAJICK, of Radiology Group of Abington signed at March 27, 2015
Document maintained by: Radiology Group of Abington

GENERAL DESCRIPTION OF THE STUDY HISTORY: Chronic liver disease.

COMPARISON: 1/22/2015

TECHNIQUE: Real time 2D grey scale and color Doppler analysis of the abdomen performed.

FINDINGS: Hepatosplenomegaly is noted. No solid mass lesions are seen. There is no intrahepatic biliary ductal dilatation.

- The gallbladder has been removed. The common bile duct is normal in caliber, measuring approximately 4 mm. Visualized portions of the pancreas are unremarkable.
- The right kidney measures 12.4 cm in long axis.
- The left kidney measures 14.3 cm in long axis.
- Bilateral simple appearing renal cysts are noted.
- A stable complex cyst in the mid-portion of the left kidney is noted.
- There is no new mass, hydronephrosis, or calculi bilaterally.
- The spleen measures 16.0 cm in long axis and is normal in echotexture. There is no ascites.
- The proximal aorta and IVC are normal in caliber.

DIAGNOSTIC IMPRESSION:
- Hepatosplenomegaly and stable renal cysts.
- No new abnormalities.

Signed by: ZAJICK, M.D., DONALD C.

9. CT Virtual Colonoscopy

Abington Health
Lansdale Hospital
Department of Radiology – Radiology Group of Abington, PC

Patient: JOHN BISOL,
Document created: September 18, 2015, 11:25 AM
Performer: MIHAELA VAVA,
Author: MIHAELA VAVA, Radiology Group of Abington
Encounter Date: September 18, 2015, 08:20 AM
Provider: DIVO MESSORI,
Information Recipient: DIVO MESSORI,
Legal Authenticator: MIHAELA VAVA, of Radiology Group of Abington
Signed: at September 18, 2015
Document maintained by: Radiology Group of Abington

GENERAL DESCRIPTION OF THE STUDY: CT virtual colonoscopy

History: Heme positive stools, colonic polyps, contraindication to colonoscopy

Comparison: CT chest, abdomen and pelvis November 2012

CT virtual colonoscopy was obtained after colonic insufflation and administration of stool tagging agent. Patient was imaged in prone and supine position. Study was review on the 3-D workstation.

Colonic insufflation was performed by Dr. Taupin.

- There is moderate distention of the ascending and transverse colon with less significant distention of the descending colon.
- Extensive sigmoid colon diverticulosis is seen and there is mild to moderate diverticulitis manifested by wall thickening and fat stranding with inflammatory changes in the left lower quadrant.
- Evaluation of the descending and sigmoid colon is limited by under distention and the presence of a mucosal abnormality at this level cannot be excluded.
- On the prone images, there is good distention of the rectum which appears unremarkable.
- Proximal sigmoid is within normal limits but the relatively large segment of sigmoid which contains multiple diverticula is not well distended which could be related to circular muscle hypertrophy.
- There is limited evaluation of the mucosa at this level.
- Lung bases are clear.
- There are extensive calcifications at the level of the aortic valve. Coronary artery calcifications are also seen.

- Liver is enlarged. Spleen is enlarged. Pancreas is grossly normal. Adrenal glands are within normal limits.
- Patient is status post cholecystectomy.
- There is a cyst in the upper pole of the left kidney measuring 36 x 23 mm with a small hyper-dense lesion in the upper pole of the left kidney measuring 13 x 16 mm which is unchanged.
- Scarring and calcification is seen at the upper pole of the right kidney as well as right renal cysts, the largest measuring 34 x 30.5 mm.
- There is a thin calcified septation in association with cyst in the lower pole of the right kidney.
- Bilateral fat containing inguinal hernias are seen.
- Bony structures show degenerative disease of the lumbar spine as well as osteophyte formation at the sacroiliac joints.
- Bilateral hip osteoarthritis is seen.

DIAGNOSTIC IMPRESSION:

- Mild sigmoid colon diverticulitis.
- No mucosal abnormality seen at the level of the ascending, transverse colon, proximal descending colon and rectum.
- Relatively long segment of the sigmoid colon which shows multiple diverticula is not well distended and shows wall thickening which could be related to circular muscle hypertrophy.
- Full evaluation of the mucosa at this level is not possible.
- Multiple incidental findings described in the body of the report.

Signed by: VAVA, M.D., MIHAELA V
Signed on: 09/18/2015 11:21:55

10. Prescription Drug List

PERMANENT – DAILY PRESCRIPTION DRUG RECORD

TRADE NAME	DOSE	INTERVAL	TOTAL DOSE
HYDRALAZINE	25mg	b.i.d	50mg
LIPITOR	40mg	q.h.s.	40mg
GABAPENTIN	300mg	q.h.s.	300mg
ALLOPURINOL	300mg	q.D	300mg
BYSTOLIC	2.5mg	q.D	2.5mg
INSULIN (TYPE N) NOVOLIN	100 Units	AM	200 Units
	100 Units	q.h.s.	
INSULIN (TYPE R) NOVOLIN	95 Units	p.c. PM	95 Units
PRAMIPEXOLE	0.250mg	q.h.s.	0.250mg
LASIX	40mg	q.D	40mg
IMODIUM	4mg	AM	4mg

DAILY SUPPLEMENTS

1 High Potency Multiple Vitamin/Mineral (Sustained Release)
1 Vitamin D-3 Supplement 1,000 Units Daily @ Bedtime
1 CoQ10 Supplement 600mg Daily

OTHER PRESCRIPTIONS

"One Touch Ultra" TEST STRIPS (LFS): Test 3 times daily
BD Insulin Syringes 1 ml 30Gauge x ½" – Use 3 times a day as directed

OTHER BOOKS BY: JOHN L. BISOL

1. The House on South Street: Second Edition	ISBN: 978-1-365-04162-4
2. Rendition I	ISBN: 978-1-4116-2220-3
3. Rendition II	ISBN: 978-0-6151-3560-1
4. BISOL – The Knight Templar	ISBN: 978-1-8472-8923-0
5. SURVIVOR – The House Sale	ISBN: 978-1-312-98902-3
6. Out of Sight – Out of Mind	ISBN: 978-1-312-94704-7
7. The House On Ernes Drive	ISBN: 978-1-312-91172-2
8. An Ostentation of Tutoring	ISBN: 978-1-329-64084-9
9. The Worcester Tornado: A New Perspective	ISBN: 978-1-329-70645-3
10. The Golden Alders	ISBN: 978-1-329-00035-3
11. An Interpretation of the Artworks: St. Francis of Assisi Church – Fitchburg, MA	ISBN: 978-1-329-71156-3
12. The House on Middle Street	ISBN: 978-1-329-77449-0
13. The House on South Street: Revisited	ISBN: 978-1-329-83038-7
14. Occupational Education: Insights & Perspectives	ISBN: 978-1-329-90631-0
15. The Veil of Cadence Shadowsoul	ISBN: 978-1-329-94140-3
16. Cleaning the House on South Street	ISBN: 978-1-365-23100-1
17. Stories of The House on South Street	ISBN: 978-1-365-24391-2
18. The Third Rendition	ISBN: 978-1-365-31587-9

How to Order:
All books are available on: amazon.com OR by ISBN @ Your Bookstore
ALSO:
http://www.lulu.com/spotlight/jlbisol
http://www.amazon.com/John-L.-Bisol/e/B00US30LJE

SPECIAL ORDER BOOK - **OBLIGATIONS: A One-Act Play**
(contact Author for License) @ www.bisol.net